T0211903

ACTION PROGRAMMING LANGUAGES

Synthesis Lectures on Artificial Intelligence and Machine Learning

Editors
Ronald J. Brachman, *Yahoo! Research*

Thomas Dietterich, *Oregon State University*

© Springer Nature Switzerland AG 2022

Reprint of original edition © Morgan & Claypool 2008

Action Programming Languages
Michael Thielscher

ISBN: 978-3-031-00419-3   paper
ISBN: 978-3-031-01547-2   ebook

DOI: 10.1007/978-3-031-01547-2

A Publication in the Springer series
SYNTHESIS LECTURES ON ARTIFICIAL INTELLIGENCE AND MACHINE LEARNING #5

Lecture #5
Series Editor: Ronald J. Brachman, Yahoo! Research and Thomas Dietterich, Oregon State University

Library of Congress Cataloging-in-Publication Data

Series ISSN: 1939-4608   print
Series ISSN: 1939-4616   electronic

ACTION PROGRAMMING LANGUAGES

Michael Thielscher
Department of Computer Science,
Technische Universität Dresden,
Dresden, Germany

SYNTHESIS LECTURES ON ARTIFICIAL INTELLIGENCE AND MACHINE LEARNING #5

ABSTRACT

Artificial systems that think and behave intelligently are one of the most exciting and challenging goals of Artificial Intelligence. *Action Programming* is the art and science of devising high-level control strategies for autonomous systems which employ a mental model of their environment and which reason about their actions as a means to achieve their goals. Applications of this programming paradigm include autonomous software agents, mobile robots with high-level reasoning capabilities, and General Game Playing. These lecture notes give an in-depth introduction to the current state-of-the-art in action programming. The main topics are

- knowledge representation for actions,
- procedural action programming,
- planning,
- agent logic programs, and
- reactive, behavior-based agents.

The only prerequisite for understanding the material in these lecture notes is some general programming experience and basic knowledge of classical first-order logic.

KEYWORDS
Agent programming, cognitive robotics, knowledge representation

Contents

CHAPTER 1

Introduction

Artificial systems that think and behave intelligently are one of the most exciting and challenging goals of Artificial Intelligence (AI). Important examples of such systems are autonomous software agents, mobile robots with high-level reasoning capabilities, and general game playing programs. All of these have in common the need for advanced cognitive functions such as the ability to follow complex and long-term strategies, to make rational decisions, to devise suitable plans, and to react sensibly in unexpected situations. These capabilities are characteristics of human-like intelligence and ultimately distinguish thinking systems from mere autonomous machines.

A fundamental paradigm in AI research says that higher intelligence is grounded in a mental representation of the world and that intelligent behavior is the result of correct reasoning with this representation. *Action Programming* is the art and science of devising high-level programs for a system which employs its own mental model to reason about its actions as a means to achieve its goals.

Research on how to design an automatic system that reasons about its actions has a long history in Artificial Intelligence (McCarthy, 1958). The classic, formal model for representing actions and for planning is the Situation Calculus, whose roots trace back to the early days of AI (McCarthy, 1963; McCarthy and Hayes, 1969). Over the years other general formalisms of similar expressiveness have been developed, most notably the Event Calculus (Kowalski and Sergot, 1986; Shanahan, 1997) and the Fluent Calculus (Thielscher, 1999). A recent variant of the Situation Calculus is the Game Description Language (Genesereth et al., 2006), a tailor-made but very expressive language that allows us to formalize the rules of arbitrary games. Other, more restricted formalisms, like the Planning Domain Definition Language (McDermott, 2000), have been introduced to admit particularly efficient reasoning systems.

Special-purpose languages not only underlie the many existing, efficient planning systems, they are also employed in agent frameworks (Georgeff and Lansky, 1987) such as AgentSpeak (Rao, 1996), SPARK (Morley and Meyers, 2004), or 3APL (Hindriks et al., 1999) and many others, which allow us to control agents on the basis of formal, symbolic world models. The more expressive Situation Calculus and similar approaches, on the other hand, provide the formal underpinnings for the general, high-level action programming languages GOLOG (Levesque *et al.*, 1997) and FLUX (Thielscher, 2005), which support the design of control programs for systems that reason about their actions and devise plans.

These lecture notes give an in-depth introduction to the current state-of-the-art in action programming. After a short recapitulation of some mathematical notions and notations in Chapter 2, we begin, in Chapter 3, with an introduction to the underlying principles for representing action knowledge that are common in a variety of diverse applications such as software agents, robots with high-level reasoning capabilities, and general game playing systems. For this representation we introduce a general action calculus, which abstracts from existing languages like the Situation Calculus or the Game Description Language. This introduction is followed by the main topic of this chapter, the procedural action programming language GOLOG, including advanced techniques such as concurrency, interrupts, and sensing.

Chapter 4 is devoted to the specific problem of planning in action programs. It is shown how efficient planning algorithms can be integrated into high-level languages like GOLOG. A special focus lies on the use of domain-specific heuristics for speeding up the planning process. This is followed by an account of recent developments concerning the problem of planning with additional preferences.

In Chapter 5 we turn to declarative action programming languages, specifically the concept of Agent Logic Programs. It is shown how the principles of logic programming can be adapted to specify agent behavior. A declarative semantics for these programs is given on the basis of the Fluent Calculus, complemented by an operational semantics along with an introduction to a programming system for declarative action programs.

Action programming for reactive, behavior-based agents are the main topic of the final Chapter 6. It contains a detailed account of the generic programming framework of Procedural Reasoning Systems (PRS), which builds on the standard BDI-model (for: Belief, Desire, Intention) for rational behavior. The programming features available in specific PRS-systems are introduced together with an account of their operational semantics.

The only prerequisite for understanding the material in these lecture notes is some general programming experience and basic knowledge of classical first-order logic. Although not essential, previous exposure to the use of logic in Artificial Intelligence may be helpful and can be gained through most standard AI textbooks, including (Nilsson, 1998) and (Russell and Norvig, 2003). While Chapter 2 includes a brief introduction to the basics of logic programming, some experience in the logic programming language Prolog will make it easier to understand the implementation details in Sections 3.5, 5.3, and 6.3. Standard textbooks in this area include (Bratko, 2000) and (Clocksin and Mellish, 2003).

CHAPTER 2

Mathematical Preliminaries

2.1 LOGIC

Classical logic will play a prominent role throughout these lecture notes. *Formulas* are built up from *atoms*, i.e., predicate symbols with terms as arguments, and the standard logical connectives, stated in order of decreasing priority: $\forall$ (*universal quantification*), $\exists$ (*existential quantification*), $\neg$ (*negation*), $\wedge$ (*conjunction*), $\vee$ (*disjunction*), $\supset$ (*implication*), and $\equiv$ (*equivalence*). Sequences of terms like $x_1, \ldots, x_n$ are often abbreviated as $\vec{x}$. Variables outside the range of the quantifiers in a formula are implicitly assumed universally quantified. We also use the *equality predicate* "=", which is always assumed to be interpreted as the identity relation.

We use a *sorted* logic language, where sorts are used to define the range of the arguments of predicates and functions. Sorts need not be disjoint; e.g., the natural numbers $\mathbb{N}$ are a sub-sort of the real numbers $\mathbb{R}$. Variables in formulas are sorted, too, and may be substituted by terms of the right sort only. For conventional sorts like the natural numbers, we use the standard arithmetic operations with their usual interpretation.

A *substitution* is a function that replaces a finite set of variables by terms, written as $\{x_1/t_1, \ldots, x_n/t_n\}$. The *application* of a substitution θ to an expression E is denoted by $E\theta$.

When logic is used to represent knowledge, it is common to make the so-called *unique name-assumption*, which says that different symbols mean different things. Formally, let $h_1, \ldots, h_n$ be a sequence of function symbols (including constants), then $UNA[h_1, \ldots, h_n]$ abbreviates the formula

$$\bigwedge_{i=1}^{n-1} \bigwedge_{j=i+1}^{n} h_i(\vec{x}) \neq h_j(\vec{y}) \ \wedge \ \bigwedge_{i=1}^{n} [h_i(\vec{x}) = h_i(\vec{y}) \supset \vec{x} = \vec{y}].$$

The first part of a unique-name axiom stipulates that terms with different leading function symbol are unequal; e.g., $Location(3) \neq Color(3)$. The second part implicitly says that terms are unequal which start with the same function symbol but whose arguments differ; e.g., $Location(3) \neq Location(4)$ given that $3 \neq 4$.

2.2   LOGIC PROGRAMMING

A *logic program* is a finite set of *clauses* of the form $p(\vec{t}) :- L_1, \ldots, L_n$, where $p(\vec{t})$ is an atom and $L_1, \ldots, L_n$ are *literals*, that is, atoms or negated atoms. If $n = 0$, then the clause is a *fact*, simply written as $p(\vec{t})$. By convention, variables in logic programs start with an uppercase letter. An example is the following set of clauses, whose purpose is to decide whether a terminal position has been reached in the well-known Tic-Tac-Toe game:

```
terminal:- line(xsymb).
terminal:- line(osymb).
terminal:- not open.


line(S):- row(S).
line(S):- col(S).
line(S):- diag(S).


row(X) :- holds(cell(M,1,X)), holds(cell(M,2,X)), holds(cell(M,3,X)).
col(X) :- holds(cell(1,N,X)), holds(cell(2,N,X)), holds(cell(3,N,X)).
diag(X):- holds(cell(1,1,X)), holds(cell(2,2,X)), holds(cell(3,3,X)).
diag(X):- holds(cell(3,1,X)), holds(cell(2,2,X)), holds(cell(1,3,X)).


open:- holds(cell(M,N,blank)).
```

The semantics of a program is to regard all clauses with the same leading predicate p as the logical definition of p. Formally, let p be any predicate symbol of the underlying signature such that

$$p(\vec{t_1}) \leftarrow L_{11}, \ldots, L_{1n_1}$$
$$\vdots$$
$$p(\vec{t_m}) \leftarrow L_{m1}, \ldots, L_{mn_m}$$

are the clauses for p in a program P ($m \geq 0$). Take a sequence $\vec{x}$ of pairwise different variables not occurring in any of the clauses, then the *logical definition* for p in P is given by the formula

$$p(\vec{x}) \equiv \bigvee_{i=1}^{m} (\exists \vec{y_i})\,(\vec{x} = \vec{t_i} \wedge L_{i1} \wedge \ldots \wedge L_{in_i})$$

where $\vec{y_i}$'s are the variables of the respective clause. The *completion* of a program P, written

COMP[*P*], is the set of definitions of all predicates in *P* along with unique-name axioms for all function symbols. For example, the completion of the Tic-Tac-Toe program from above is, after some straightforward simplification of equalities,

$$Terminal \equiv Line(XSymb) \lor Line(OSymb) \lor \neg Open$$
$$Line(s) \equiv Row(s) \lor Col(s) \lor Diag(s)$$
$$Row(x) \equiv (\exists m) \ (Holds(Cell(m, 1, x)) \land Holds(Cell(m, 2, x)) \land Holds(Cell(m, 3, x)))$$
$$Col(x) \equiv (\exists n) \ (Holds(Cell(1, n, x)) \land Holds(Cell(2, n, x)) \land Holds(Cell(3, n, x)))$$
$$Diag(x) \equiv Holds(Cell(1, 1, x)) \land Holds(Cell(2, 2, x)) \land Holds(Cell(3, 3, x))$$
$$\lor$$
$$Holds(Cell(3, 1, x)) \land Holds(Cell(2, 2, x)) \land Holds(Cell(1, 3, x))$$
$$Open \equiv (\exists m, n) \ Holds(Cell(m, n, Blank))$$

A *query* Q to a logic program P is a sequence of literals. It encodes the question whether $(\exists \vec{x})Q$ is a logical consequence from *COMP*[*P*]. Queries are computed by *derivations*, by which a query is successively rewritten. A single derivation step for a query $Q = L_1, L_2, \ldots, L_n$ produces a new query as follows:

- Suppose L_1 is a positive atom. Let $C :\text{-} C_1, \ldots, C_m$ be a clause in P (with all variables renamed) for which there exists a substitution θ such that $L_1\theta = C\theta$. The application of this clause results in the new query $Q' = (C_1, \ldots, C_m, L_2, \ldots, L_n)\theta$.
- Suppose L_1 is a negative literal $\neg A$. If no successful derivation for A itself exists, then the new query is simply $Q' = L_2, \ldots, L_n$ (the *negation as failure*-principle).

A *successful* derivation ends with the empty query, denoted by $\Box$. In a successful derivation, the substitutions used in each derivation step can be combined and then restricted to the variables in the original queries. The resulting substitution is called the *computed answer*. This computation mechanism is known to be semantically correct, that is, if θ is a computed answer for a query Q given program P, then $Q\theta$ is a logical consequence of the program, written $COMP[P] \models Q\theta$.

As an example, suppose the program for Tic-Tac-Toe is augmented by the following encoding of a concrete position:

```
holds(cell(1,1,xsymb)). holds(cell(1,2,osymb)). holds(cell(1,3,xsymb)).
holds(cell(2,1,blank)). holds(cell(2,2,osymb)). holds(cell(2,3,xsymb)).
holds(cell(3,1,osymb)). holds(cell(3,2,blank)). holds(cell(3,3,xsymb)).
```

In this situation there is a successful derivation for the query `terminal`:

terminal
----
line(xsymb)
----
col(xsymb)
----
holds(cell(1,N,xsymb)), holds(cell(2,N,xsymb)), holds(cell(3,N,xsymb))
----
holds(cell(2,3,xsymb)), holds(cell(3,3,xsymb))
----
holds(cell(3,3,xsymb))

□

CHAPTER 3

Procedural Action Programs

3.1 SETTING THE STAGE: DEFINING FLUENTS AND ACTIONS

A standard computer program is mainly concerned with the manipulation of variables. The variables are defined prior to their use, they get an initial value, and a basic statement in a computer program is the assignment of a new value to a variable. The values of all variables together at some point during the execution of a program form an internal *state*. A new variable assignment then corresponds to a state transition, and the execution of a program results in a sequence of transitions from the initial state, i.e., the initial variable settings, to some terminal state.

Action programs differ from standard programs in that the basic statements can be arbitrary actions, which are to be performed by an agent in its environment. These actions typically affect the environment, and the goal of an action program is to generate an action sequence that brings the initial state of the environment into some desired goal state. In order to achieve this, the relevant properties of the environment need to be symbolically represented in an action program. The definition of these properties, which are traditionally called *fluents*, is necessary for every action program and corresponds to the declaration of variables in standard programs.

In action programs, a single *action* denotes a specific way of interacting with the environment. Actions may change the outside world, e.g., when a robot picks up an object or a software agent orders a product over the Internet. Other actions only change the status of the physical agent itself, e.g., when a robot moves to a new position. Finally, actions may just provide the agent with information about the environment, e.g., when a robot senses whether a door is open or a software agent compares prices at different online stores. While a single action can be a very complex behavior on the level of the physical agent, actions are taken as elementary entities on the level of action programs.

Example: Mail Delivery Robot

Imagine a robot sitting in a hallway with a number of offices in a row. The robot is an automatic post boy, whose task is to pick up and deliver packages exchanged among the offices. Figure 3.1 depicts a particular scenario in an environment with six offices, a robot that can carry at most three packages at a time, and nine packages waiting for delivery. The action programming task

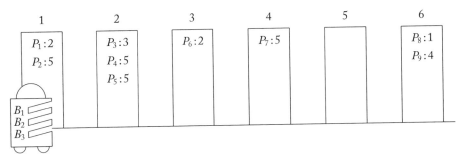

FIGURE 3.1: The initial state of a package delivery problem. All three mail bags of the robot are empty, and there are currently nine packages waiting for delivery. The associated number indicates their destination.

here is to write a control program that sends the robot up and down the hallway and tells it where to collect and drop packages.

Assuming that people in the offices can issue an unbounded number of delivery requests, the environment can be in any of an unbounded number of different states. To encode these states, the following parameters—fluents—shall be used, where PACKAGE shall be the set of package identifiers, ROOM $= \{1, \ldots, 6\}$, and BAG $= \{B_1, B_2, B_3\}$ the three mail bags:

Symbol	Type
At	ROOM $\mapsto$ FLUENT
Empty	BAG $\mapsto$ FLUENT
Carries	BAG $\times$ PACKAGE $\times$ ROOM $\mapsto$ FLUENT
Request	PACKAGE $\times$ ROOM $\times$ ROOM $\mapsto$ FLUENT

All these fluents are *relational*, that is, they are either true or false in a state of the environment. Specifically, $At(r)$ is true if the robot is at room r, $Empty(b)$ means that the robot's mail bag b is empty, $Carries(b, p, r)$ indicates that bag b carries package p with destination r, and $Request(p, r_1, r_2)$ denotes the request to deliver package p from room r_1 to r_2. All four expressions are terms of a special sort FLUENT.

The robot can manipulate the state variables with the help of the following symbolic actions, which will constitute the basic statements of the action program for the robot:

Symbol	Type
Go	DIRECTION $\mapsto$ ACTION
Pick	PACKAGE $\times$ BAG $\mapsto$ ACTION
Drop	BAG $\mapsto$ ACTION

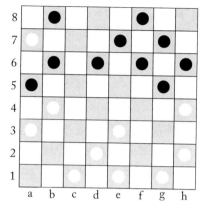

FIGURE 3.2: On a cylindrical checkerboard, the a-file borders on the h-file. Therefore, the white piece now on a7 could move to h8 to promote. If it is Black's move, then the piece on a5 could promote on a1 by a double jump via g3, thereby capturing the white pieces on h4 and h2.

Suppose DIRECTION $= \{Up, Down\}$, then action $Go(d)$ means to move up or down the hallway to the next office; action $Pick(p, b)$ asks the robot to pick up package p and put it in bag b; and action $Drop(b)$ denotes the delivery of the contents of bag b at the current location.

Example: Cylindrical Checkers

A *General Game Player* is a system that understands the formal description of an arbitrary game and learns to play this game well without human intervention. As an example, let us consider a slight variant of the well-known game of Checkers with the standard rules but played on a cylindrical board; cf. Figure 3.2. The rules of this game can be formalized on the basis of just two fluent symbols, which can be combined into any of the possible board positions:

Symbol	Type	Range
Cell	FILE × ROW ↦ FLUENT	{*Blank, White, WhiteKing, Black, BlackKing*}
Control	↦ FLUENT	{*White, Black*}

Both these fluents are *functional*, that is, in every state of the environment they take any of the values from the given range. Specifically, *Cell*(x, y) is the contents of a particular cell, and the value of *Control* is the party whose move it is in the current position. The various possible moves of the two players in Checkers can be encoded by a single action:

Symbol	Type
Move	FILE × ROW × FILE × ROW ↦ ACTION

An instance $Move(x_1, y_1, x_2, y_2)$ may be either a straight move or a jump of either a piece or a king from square (x_1, y_1) to (x_2, y_2). Double and triple jumps etc. can be encoded as sequences of single jumps.

3.2   GOLOG PROGRAMS

Having defined the fluents and primitive actions at the disposal of an agent, action programming is the art of designing complex behaviors out of the basic actions. The action programming language GOLOG, an acronym for *Algol in Logic*, combines actions and fluents with standard constructs from procedural programming languages. A GOLOG program itself is a sequence of definitions for any number of procedures $p_1(\vec{v}_1), \ldots, p_n(\vec{v}_n)$ followed by a main body:

$$\textbf{proc } p_1(\vec{v}_1)\, \delta_1 \textbf{ endProc}; \ldots; \textbf{ proc } p_n(\vec{v}_n)\, \delta_n \textbf{ endProc}; \delta$$

where $n \geq 0$. The main body δ as well as the body δ_i of a procedural definition is built on the syntactic elements shown in Figure 3.3. The basic commands in every GOLOG program are the primitive actions of the underlying domain. Conditions (tests) are based on the fluents defined in the domain. GOLOG has several nondeterministic features, whose semantics will be made precise in the following section. First, however, let us have a look at an example program, which illustrates the typical use of the various programming constructs.

Command	Meaning
nil	empty program
a	primitive action
ϕ ?	test
$\delta_1 ; \delta_2$	sequential composition
$\delta_1 \mid \delta_2$	nondeterministic choice of sub-program
$\pi\, x . \delta(x)$	nondeterministic choice of argument
δ^*	nondeterministic iteration
$p(\vec{t})$	| procedure call
if ϕ **then** δ_1 **else** δ_2 **endIf**	conditional
while ϕ **do** δ **endWhile**	loop

FIGURE 3.3: The syntactic elements of GOLOG programs. The expression ϕ is a logic formula based on the fluents of the domain. The sub-programs $\delta, \delta_1, \delta_2$ are recursively defined using all programming constructs.

A GOLOG Program for Mail Delivery

It is quite straightforward to come up with a simple, effective strategy for the mail delivery problem. As long as the robot finds itself at some office for which it carries one or more packages, it should select these packages, in no particular order, for immediate delivery. Conversely, if the robot happens to be at some place where packages are waiting to be collected and the mailbags are not yet full, then nondeterministically one package after the other gets chosen and put into one of the empty mailbags. If, however, no more packages can be dropped nor collected at its current location, the robot makes an arbitrary decision to move either up or down the hallway toward some office for which it has mail or where packages are still waiting.

Using the fluents and actions defined earlier, this algorithm can be implemented by a GOLOG program as follows. To begin with, consider these two simple procedure definitions:

proc *Deliver*
  $\pi b. Drop(b)$
endProc;

proc *Collect*
  $\pi b. \pi p. Pick(p, b)$
endProc;

The first procedure nondeterministically selects a bag and leaves its contents at the current location. As will be shown in Section 3.3, the execution model for GOLOG ensures that only those arguments b are selected for which the action is actually possible in the current state of the environment. In a similar fashion, the second procedure selects a package p to be picked up and put in one of the mail bags.

The third procedure to be used in the control program handles the case where the delivery robot has to decide to move either up or down the hallway to the next office:

proc *Continue*
  if $(\exists b) \neg Empty(b)$ **then**
    $\pi b. \pi p. \pi r. \pi r'. (At(r)?; Carries(b, p, r')?;$
                $\text{if } r < r' \text{ then } Go(Up) \text{ else } Go(Down) \text{ endIf})$
  else
    $\pi p. \pi r. \pi r_1. \pi r_2. (At(r)?; Request(p, r_1, r_2)?;$
                $\text{if } r < r_1 \text{ then } Go(Up) \text{ else } Go(Down) \text{ endIf})$
  endIf
endProc;

The intuition is as follows: if there is a non-empty mail bag, then a room is (nondeterministically) chosen for which a package is being carried, and the robot moves up or down the hallway depending on whether this room is to the left or right of the current location. More specifically, values b, p, r, and r' are selected for a bag, a package, the current location, and the destination, respectively, so that the tests $At(r)$? and $Carries(b, p, r')$? succeed, and then the choice of the Go action to be performed depends on whether or not $r < r'$. If, on the other hand, all mail bags are empty, then one of the remaining requests is nondeterministically chosen and the robot moves toward the location where the respective package can be picked up.

The following main control procedure completes the GOLOG program for the mail delivery problem:

> **proc** *Control*
>   **while** $(\exists b) \neg Empty(b) \vee (\exists p, r_1, r_2)\, Request(p, r_1, r_2)$ **do**
>   **if** $(\exists b, p, r)\,(Carries(b, p, r) \wedge At(r))$ **then**
>   *Deliver*
>   **else**
>   **if** $(\exists b, p, r_1, r_2)\,(Empty(b) \wedge Request(p, r_1, r_2) \wedge At(r_1))$ **then**
>   *Collect*
>   **else**
>   *Continue*
>   **endIf**
>   **endIf**
>   **endWhile**
> **endProc;**

The main body of the program shall be a simple call to this control procedure, so that the entire GOLOG program is

> **proc** *Deliver* ... **proc** *Control* ... **endProc;** *Control*

Following this strategy, our robot in Figure 3.1 would pick up the two packages in the first room and then move up to room number 2, where it delivers P_1. Thereafter, it selects two of the three packages waiting there and continues to move to the right. Provided that no further request is issued during the execution of this program, it eventually terminates with all delivery requests satisfied.

GOLOG programs that make use of nondeterminism may admit various execution traces. This gives rise to two conceptually different modes of execution: *online execution* means to run the program step-by-step, and whenever this requires to do a primitive action, then this action is carried out immediately by the physical agent. This is the standard way of implementing control programs on physical agents. Running a program in this mode means to commit to every nondeterministic selection, because actions cannot just be taken back once they have been performed by the agent. This may have the disadvantage that the chosen course of actions eventually leads to a dead-end, that is, where the program cannot be completed although a different run would have been successful.

In *offline execution*, on the other hand, a program is first run in simulation. This allows us to check several, possibly all, ways to finish a program prior to having an agent commit to a particular trace. In this way, it is guaranteed that a successful execution will be found whenever one exists. A further advantage of this execution principle is that it allows us to find the most economical way of running a program (say, in terms of the number of primitive actions to be performed) prior to having the agent actually perform actions. The disadvantage, however, is that the longer a program the more possible execution paths it tends to have, so that it is practically impossible to check them all. Moreover, pure offline execution is not applicable if the agent has to use its sensing capabilities to acquire important information at runtime and which is available only after parts of the program have been executed. A sequence of actions generated by offline execution may also become invalid if exogenous actions occur, that is, actions besides those performed by the agent. A mixture of both execution modes combines the best of both worlds—more to this in Section 3.7.

No matter which execution mode is chosen, running an action program requires to evaluate conditions which depend on the current state of the environment in which the agent lives. Since these properties frequently change as the program proceeds and not all of them may be directly observable by the agent, executing a program requires to maintain an internal model of the environment, which throughout the execution of the program conveys the necessary information about the relevant fluents. The model needs to be updated after each action in accordance with the effects of the action. The execution of a GOLOG program therefore relies on an *action knowledge base*, in which the preconditions and effects of the actions are specified. The aim of *action calculi* is to use classical logic to axiomatize actions and their effects.

3.3   ACTION CALCULI

3.3.1   Signatures with Relational Fluents

As we have seen, fluents and actions are the fundamental domain-dependent ingredients of every action program. Therefore, fluents and actions are the basic sorts in the sorted logic language we are going to define. Action calculi also need to distinguish different points in time in order to axiomatize the changes caused by actions. As a third fundamental sort, we therefore assume an abstract notion of time. A simple example of a time structure are the natural numbers, which model a linear, discrete time line. More complex notions of time involve continuous change (e.g., modeled by the positive rationals) or a branching time structure to denote different potential evolutions of the environment.

The three basic sorts are used for three fundamental predicates. The relation $t_1 < t_2$ denotes a (possibly partial) ordering on the time structure. Predicate $Holds(f, t)$ is used to say that fluent f is true at time t. Finally, the intended meaning of expression $Poss(a, s, t)$ is that it is possible to do action a beginning at time s and ending at time t. These predicates, along with the three fundamental sorts, form the basis of a domain signature in a general action calculus.

Definition 3.3.1. *A domain signature is a finite, sorted logic language which includes the sorts* FLUENT, ACTION, *and* TIME *along with the predicates*

$$<: \text{TIME} \times \text{TIME}$$
$$Holds: \text{FLUENT} \times \text{TIME}$$
$$Poss: \text{ACTION} \times \text{TIME} \times \text{TIME}$$

As usual, $s \leq t$ stands for $s < t \vee s = t$.   □

Throughout the lecture notes we denote variables of sort ACTION by the letter a, variables of sort FLUENT by f and g, and variables of sort TIME by s and t. We tacitly assume *uniqueness-of-names* for all fluents and actions. That is to say, different fluent terms denote different state properties, and different action terms denote different things, too.

Next, we define the notion of a state formula, which allows us to express properties of a domain at given times.

Definition 3.3.2. *Let $\vec{t}$ be a non-empty sequence of variables of sort* TIME *in a given domain signature. A* state formula *in $\vec{t}$ is a first-order formula $\Phi[\vec{t}]$ in which the variables in $\vec{t}$ occur free and such that*

- *for each occurrence of $Holds(f, t)$ in Φ we have $t \in \vec{t}$;*
- *predicate $Poss$ does not occur in Φ.*   □

We are now in a position to define, in our general calculus, the two basic elements of a knowledge base for actions to be used as the background theory for an action program: precondition axioms, which define the conditions for actions to be applicable, and effect axioms, which define the consequences of actions. For the latter, we use a general form that allows us to define different "cases" $i = 1, \ldots, k$ of updates $\Upsilon_i[s, t]$ (cf. axiom (3.1)). Each of these sub-formulas defines the fluents that hold after the action, at time t, relative to the state when the action starts, at time s.

Definition 3.3.3. *Consider a domain signature, and let A be a function into sort* ACTION.

- *A* precondition axiom *is of the form*

$$Poss(A(\vec{x}), s, t) \equiv \pi_A[s]$$

  where $\pi_A[s]$ is a state formula in s with free variables among s, t, $\vec{x}$.

- *An* effect axiom *is of the form*

$$Poss(A(\vec{x}), s, t) \supset \Upsilon_1[s, t] \vee \ldots \vee \Upsilon_k[s, t] \tag{3.1}$$

  where $k \geq 1$ and each $\Upsilon_i[s, t]$ $(1 \leq i \leq k)$ is a formula of the form

$$
\begin{aligned}
(\exists \vec{y}_i)(\Phi_i[s] \;\wedge\; (\forall f)\,[\Gamma_i^+[s, t] \supset Holds(f, t)] \\
\wedge\; (\forall f)\,[\Gamma_i^-[s, t] \supset \neg Holds(f, t)])
\end{aligned}
\tag{3.2}
$$

  in which $\Phi_i[s]$ is a state formula in s with free variables among s, $\vec{x}$, $\vec{y}_i$,[1] and both $\Gamma_i^+[s, t]$ and $\Gamma_i^-[s, t]$ are state formulas in s, t with free variables among f, s, t, $\vec{x}$, $\vec{y}_i$.

A domain axiomatization *consists of precondition and effect axioms, one each for every function into sort* ACTION. □

Prior to being treated to some example axioms, recall that the purpose of an action program is to trigger the agent to do the right action at the right time. Every agent program computes a sequence of actions to be executed by the agent. In order to distinguish the different possible executions of a program, it is convenient to resort to a *branching* time structure, based on the concept of a *situation*. Lending the Situation Calculus its name, a situation denotes the sequence of actions that have been performed up to a certain stage of an actual program run. The special constant S_0 denotes the initial situation at the beginning of a program, when no primitive action has yet been performed. The constructor $Do(a, s)$ then maps an action a and a situation s to the situation after the performance of the action. Hence, action sequences are nested terms of the form $Do(a_n, \ldots, Do(a_1, S_0)\ldots)$. The situations can be visualized as the nodes of a tree rooted in S_0; see Figure 3.4. Each branch in this tree is a potential run of a program for the agent.

Example: Mail Delivery Axioms

Based on the branching time structure, the preconditions for the three actions in the mail delivery world can be formalized as follows. Going up the hallway is possible unless the robot happens to be at the upper end of the hallway; in other words, the actions is possible

---

[1]The purpose of sub-formula $\Phi_i[s]$ is to define possible restrictions for case i to apply.

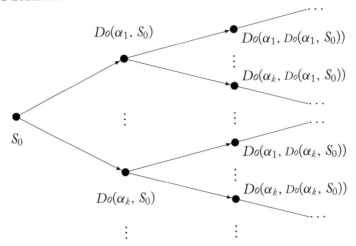

FIGURE 3.4: A tree of situations.

from situation s to situation t whenever the robot is at a room $r < 6$ in situation s and $t = Do(Go(d), s)$ with $d = Up$. Similarly, to be able to go down the robot must be at some office $r > 1$. Picking up a package p and putting it into mail bag b is possible in a situation in which bag b is empty and the robot is at room r such that there is the request to take package p from r to some r_1. Finally, the contents of a bag b can be dropped whenever the robot carries a package in b and this package is for the room where the robot currently is. This is summarized in the following precondition axioms:

$$Poss(Go(d), s, t) \equiv t = Do(Go(d), s) \wedge$$
$$(\exists r)(Holds(At(r), s) \wedge [d = Up \wedge r < 6 \vee d = Down \wedge r > 1])$$

$$Poss(Pick(p, b), s, t) \equiv t = Do(Pick(p, b), s) \wedge$$
$$(\exists r, r_1)(Holds(At(r), s) \wedge Holds(Request(p, r, r_1), s) \wedge \qquad (3.3)$$
$$Holds(Empty(b), s))$$

$$Poss(Drop(b), s, t) \equiv t = Do(Drop(b), s) \wedge$$
$$(\exists p, r)(Holds(At(r), s) \wedge Holds(Carries(b, p, r), s))$$

As an example, consider the formal description of the initial situation depicted in Figure 3.1:

$$Holds(f, S_0) \equiv f = At(1) \vee f = Empty(B_1) \vee f = Empty(B_2) \vee f = Empty(B_3) \vee$$
$$f = Request(P_1, 1, 2) \vee f = Request(P_2, 1, 5) \vee \ldots \vee \qquad (3.4)$$
$$f = Request(P_8, 6, 1) \vee f = Request(P_9, 6, 4)$$

The precondition axioms then imply, for instance, that $Poss(Go(Up), S_0, Do(Go(Up), S_0))$ but not $Poss(Go(Down), S_0, t)$ for any t. Moreover, with two packages at the current location and all three mail bags empty, there are six possible ways of picking up a package, e.g., $Poss(Pick(P_1, B_3), S_0, Do(Pick(P_1, B_3), S_0))$.

The effects of the three actions are formalized as follows. To begin with, action $Go(d)$ has two alternative effects, depending on whether $d = Up$ or $d = Down$. In the former case, the only positive effect is that fluent $At(r + 1)$ becomes true, where r is the current location of the robot. Likewise, if the robot moves down the hallway, then the only positive effect is that $At(r - 1)$ becomes true. The only negative effect in both cases is that $At(r)$ itself becomes false. All other fluents keep their truth-value from s to t. This is summarized in the following effect axiom:

$$
\begin{aligned}
&Poss(Go(d), s, t) \supset \\
&\quad d = Up \wedge (\exists r)(Holds(At(r), s) \wedge \\
&\qquad (\forall f)[f = At(r + 1) \vee (Holds(f, s) \wedge f \neq At(r)) \supset Holds(f, t)] \\
&\qquad \wedge \\
&\qquad (\forall f)[f = At(r) \vee (\neg Holds(f, s) \wedge f \neq At(r + 1)) \supset \neg Holds(f, t)]) \\
&\quad \vee \\
&\quad d = Down \wedge (\exists r)(Holds(At(r), s) \wedge \\
&\qquad (\forall f)[f = At(r - 1) \vee (Holds(f, s) \wedge f \neq At(r)) \supset Holds(f, t)] \\
&\qquad \wedge \\
&\qquad (\forall f)[f = At(r) \vee (\neg Holds(f, s) \wedge f \neq At(r - 1)) \supset \neg Holds(f, t)])
\end{aligned}
\tag{3.5}
$$

For instance, given that $Poss(Go(Up), S_0, S_1)$, where $S_1 = Do(Go(Up), S_0)$, the effect axiom entails $Holds(At(2), S_1)$ given that $Holds(At(1), S_0)$. Also, $\neg Holds(At(3), S_1)$ since $\neg Holds(At(3), S_0)$ and $At(3) \neq At(2)$, etc.

Turning to the action of putting a package into a mail bag, the only positive effect is that the package in question is now being carried. A negative effect is that the bag is no longer empty, and a further negative effect shall be that the corresponding request is no longer present. Formally,

$$
\begin{aligned}
&Poss(Pick(p, b), s, t) \supset \\
&\quad (\exists r_1, r_2)(Holds(Request(p, r_1, r_2), s) \wedge \\
&\qquad (\forall f)[f = Carries(b, p, r_2) \vee (Holds(f, s) \wedge f \neq Empty(b) \wedge f \neq Request(p, r_1, r_2)) \\
&\qquad\quad \supset Holds(f, t)] \\
&\qquad \wedge \\
&\qquad (\forall f)[f = Empty(b) \vee f = Request(p, r_1, r_2) \vee (\neg Holds(f, s) \wedge f \neq Carries(b, p, r_2)) \\
&\qquad\quad \supset \neg Holds(f, t)])
\end{aligned}
\tag{3.6}
$$

Considering initial situation (3.4) again, we have already seen that $Poss(Pick(P_1, B_3), S_0, S_1)$ with $S_1 = Do(Pick(P_1, B_3), S_0)$. The effect axiom then entails all that can be concluded about the resulting situation, that is,

$$Holds(f, S_1) \equiv f = At(1) \vee f = Empty(B_1) \vee f = Empty(B_2) \vee f = Carries(B_3, P_1, 2) \vee$$
$$f = Request(P_2, 1, 5) \vee \ldots \vee f = Request(P_8, 6, 1) \vee f = Request(P_9, 6, 4)$$

Hence, bag B_3 now carries the package in question and is no longer empty, while the request for P_1 has disappeared and all other fluents remain unchanged.

To complete this example, here is a suitable specification of the effects of dropping the contents of a mail bag:

$$Poss(Drop(b), s, t) \supset$$
$$(\exists p, r)\, (Holds(Carries(b, p, r), s) \wedge$$
$$(\forall f)\, [f = Empty(b) \vee (Holds(f, s) \wedge f \neq Carries(b, p, r)) \supset Holds(f, t)] \qquad (3.7)$$
$$\wedge$$
$$(\forall f)\, [f = Carries(b, p, r) \vee (\neg Holds(f, s) \wedge f \neq Empty(b)) \supset \neg Holds(f, t)])$$

With the help of a precondition and effect axiom for each of the three actions, and given the specification of an initial situation, it is possible to infer the executability of arbitrary action sequences and to compute the overall resulting situation.

3.3.2   Signatures with Functional Fluents

It is often more compact to use functional fluents instead of relational ones. While the latter are binary (because they are either true or false in a state), the values of a functional fluent are taken from an arbitrarily chosen range. This requires a slightly modified definition of a domain signature, where the predicate $Holds(f, t)$ is replaced by the function $Val(f, t)$, indicating the value of a fluent at time t.

Definition 3.3.4. *A functional domain signature is as in Definition 3.3.1 but with an additional sort* VALUE *and where predicate Holds is replaced by the function*

$$Val : \text{FLUENT} \times \text{TIME} \mapsto \text{VALUE}$$

Consider a functional domain signature, and let $\vec{t}$ be a non-empty sequence of variables of sort TIME *and A be a function into sort* ACTION.

- *A state formula in $\vec{t}$ is a first-order formula $\Phi[\vec{t}\,]$ in which the variables in $\vec{t}$ occur free and such that*
  - *for each occurrence of Val(f, t) in Φ we have $t \in \vec{t}$;*
  - *predicate Poss does not occur in Φ.*

- *A* precondition axiom *is of the form*

$$Poss(A(\vec{x}), s, t) \equiv \pi_A[s]$$

  where $\pi_A[s]$ is a state formula in s with free variables among s, t, $\vec{x}$.
- *An* effect axiom *is of the form*

$$Poss(A(\vec{x}), s, t) \supset \Upsilon_1[s, t] \vee \ldots \vee \Upsilon_k[s, t] \tag{3.8}$$

  where $k \geq 1$ and each $\Upsilon_i[s, t]$ $(1 \leq i \leq k)$ is a formula of the form

$$(\exists \vec{y}_i)(\Phi_i[s] \wedge (\forall f, v)[\Gamma_i[s, t] \supset Val(f, t) = v]) \tag{3.9}$$

  in which $\Phi_i[s]$ is a state formula in s with free variables among s, $\vec{x}$, $\vec{y}_i$, and $\Gamma_i[s, t]$ is a state formula in s, t with free variables among f, s, t, $\vec{x}$, $\vec{y}_i$, v. □

Example: Axiomatizing Checkers

The rules of arbitrary games can be formalized with the help of precondition and effect axioms for the possible moves. As an example, the conditions for a legal move in the cylindrical variant of Checkers may look like this—where, for the sake of simplicity, we resort to a linear and discrete time structure and use an auxiliary predicate to specify the actual occurrence of an action:

$$Poss(Move(x_1, y_1, x_2, y_2), s, t) \equiv$$
$$Occurs(Move(x_1, y_1, x_2, y_2), s) \wedge t = s + 1 \wedge$$
$$[LegalWhiteMove \vee LegalBlackMove \vee LegalKingMove \vee LegalJump]$$

Here, each of the four disjuncts encodes the conditions for one of the possible legal moves, so that, say, *LegalWhiteMove* is

$$Val(Control, s) = White \wedge Val(Cell(x_1, y_1), s) = White \wedge Val(Cell(x_2, y_2), s) = Blank \wedge$$
$$y_1 < 8 \wedge y_2 = y_1 + 1 \wedge NeighborFiles(x_1, x_2)$$

The auxiliary relation *NeighborFiles* allows us to distinguish cylindrical Checkers from the standard variant by including *NeighborFiles(a, h)* and *NeighborFiles(h, a)* in its definition. The other sub-formulas in the precondition axiom can be likewise specified according to the laws of Checkers; the details shall be omitted here.

The position change caused by a move in Checkers can be axiomatized by the following effect axiom:

$$Poss(Move(x_1, y_1, x_2, y_2), s, t) \supset LegalWhiteMove \wedge WhiteMoveUpdate \vee$$
$$LegalBlackMove \wedge BlackMoveUpdate \vee$$
$$LegalKingMove \wedge KingMoveUpdate \vee$$
$$LegalJump \wedge JumpUpdate$$

where, for example, *WhiteMoveUpdate* abbreviates the sub-formula

$$(\forall f, v)[\ f = Cell(x_1, y_1) \wedge v = Blank \vee$$
$$f = Cell(x_2, y_2) \wedge ((y_2 < 8 \wedge v = White) \vee (y_2 = 8 \wedge v = WhiteKing)) \vee$$
$$f = Control \wedge v = Black \vee$$
$$f \neq Cell(x_1, y_1) \wedge f \neq Cell(x_2, y_2) \wedge f \neq Control \wedge v = Val(f, s)$$
$$\supset Val(f, t) = v]$$

Put in words, moving a white piece has the effect of cell (x_1, y_1) becoming blank, cell (x_2, y_2) housing a white piece or a white king (in case of a promotion on the 8th row), and control going to Black. All other fluent values remain unchanged. The remaining cases of legal moves can be formulated in a straightforward fashion; again we omit the details.

3.4   GOLOG SEMANTICS

The underlying knowledge base for the elementary actions is needed to execute a GOLOG program. Specifically, because conditionals are evaluated against the current state of the environment, the internal world model must at any time correctly reflect this state. A *successful* run of a program determines a particular sequence of primitive actions being executed by the agent. A specific run therefore corresponds to a particular branch in the situation tree, determining a final situation. Consequently, the semantics of a GOLOG program is given by a characterization of the situations that correspond to a successful execution of the program. This is formally expressed by a relation $DO(\delta, s, s')$ with the intended reading that starting in situation s, the (sub-)program δ can be successfully executed ending in situation s'. This relation is inductively defined over the various programming constructs as follows:[2]

$$\begin{aligned} DO(\mathbf{nil}, s, s') &\overset{\text{def}}{=} s' = s & \text{empty program} \\ DO(a, s, s') &\overset{\text{def}}{=} Poss(a, s, s') \wedge s' = Do(a, s) & \text{primitive action} \\ DO(\phi?, s, s') &\overset{\text{def}}{=} \phi[s] \wedge s' = s & \text{test} \\ DO(\delta_1; \delta_2, s, s') &\overset{\text{def}}{=} (\exists s'')(DO(\delta_1, s, s'') \wedge DO(\delta_2, s'', s')) & \text{sequence} \\ DO(\delta_1 | \delta_2, s, s') &\overset{\text{def}}{=} DO(\delta_1, s, s') \vee DO(\delta_2, s, s') & \text{choice 1} \\ DO(\pi x.\delta(x), s, s') &\overset{\text{def}}{=} (\exists x) DO(\delta(x), s, s') & \text{choice 2} \end{aligned}$$

[2]For the sake of simplicity, we refrain from stating the precise definition for procedure calls, as this requires more involved a semantics in case of recursive procedures.

The expression $\phi[s]$ in the definition for a test stands for formula ϕ with all occurrences of a fluent f replaced by the atom $Holds(f, s)$ (or $Val(f, s)$ in case of functional fluents). These definitions are straightforward given the intuitive meaning of the various programming constructs. The semantics for nondeterministic iteration is more involved as it requires transitive closure, which can only be defined by a second-order logic formula.

Second-order logic adds sorted variables for predicates and functions to the language of first-order logic. Interpretations for second-order formulas assign relations (of the right arity and sort) to predicate variables, and mappings (of the right arity and sort) to function variables. The second-order formula $(\forall P)(\exists x)\, P(x)$, for example, is unsatisfiable, because there exists an assignment for variable P, namely, the empty relation, which is false for any x. Substitutions for predicate and function variables in formulas use *λ-expressions*. These are of the form $\lambda x_1, \ldots, x_n.\tau$ with n being the arity of the variable and where τ is a first-order formula or term, respectively. The result of the substitution is that the variable expression $P(t_1, \ldots, t_n)$ (or $f(t_1, \ldots, t_n)$, respectively) is replaced by $\tau\{x_1/t_1, \ldots, x_n/t_n\}$. For example, applying the substitution $\{P/\lambda x.\ Holds(Empty(B_1), x) \wedge \neg Holds(Empty(B_1), x)\}$ to the formula $(\exists s)\, P(s)$ results in the (inconsistent) first-order formula $(\exists s)\,(Holds(Empty(B_1), s) \wedge \neg Holds(Empty(B_1), s))$.

With the help of second-order quantification, the semantics of nondeterministic iteration can be given by the following formula:

$$DO(\delta^*, s, s') \overset{\text{def}}{=} (\forall P)\ ([(\forall s_1)\, P(s_1, s_1) \wedge (\forall s_1, s_2, s_3)\,(P(s_1, s_2) \wedge DO(\delta, s_2, s_3) \supset P(s_1, s_3))]$$
$$\supset\ P(s, s'))$$

Put in words, executing δ zero or more times takes one from situation s to situation s' just in case (s, s') is in every set (hence, the smallest set) such that

- (s_1, s_1) is in the set for all situations s_1;
- whenever (s_1, s_2) is in the set and doing δ in situation s_2 takes one to s_3, then (s_1, s_3) is in the set.

The last two missing commands in GOLOG programs, conditionals and loops, can be expressed as mere abbreviations using the other constructs:

$$\textbf{if } \phi \textbf{ then } \delta_1 \textbf{ else } \delta_2 \textbf{ endIf } \overset{\text{def}}{=} (\phi\,?;\, \delta_1)\,|\,(\neg\phi\,?;\, \delta_2)$$
$$\textbf{while } \phi \textbf{ do } \delta \textbf{ endWhile } \overset{\text{def}}{=} (\phi\,?;\, \delta)^*;\, \neg\phi\,?$$

(3.10)

As an example, recall the procedure *Collect* in the mail delivery program. Its semantics is given as

$$DO(Collect, s, s') \stackrel{\text{def}}{=} DO(\pi b.\pi p.Pick(p, b), s, s')$$
$$\stackrel{\text{def}}{=} (\exists b)(\exists p)(Poss(Pick(p, b), s, s') \wedge s' = Do(Pick(p, b), s))$$

Hence, given initial situation (3.4) along with the underlying domain axiomatization, there are a total of six possible executions of this procedure, including $s = Do(Pick(P_1, B_3), S_0)$. This is also one of the six elementary actions that can start the execution of the main body of the GOLOG program (cf. page 12): the while-condition is true in S_0, the first if-condition is false, but the second if-condition holds. Due to the nondeterministic features, there are many successful executions of the program, each corresponding to a sequence of primitive actions after which all requests have been carried out. All of these runs include nine instances of *Pick* and *Drop* actions, but they may differ in the number of *Go* actions. The shortest of the possible resulting situations has a total number of 32 primitive actions, the longest needs 34 actions.

3.5 A GOLOG INTERPRETER

This section gives a brief introduction to the use of logic programming to implement an interpreter for GOLOG programs. This includes a method to specify a background theory as part of the logic program. In accordance with the branching time structure used in the semantics for GOLOG, the domain axiomatization is based on the concept of a situation. In this setting, also known as the *Situation Calculus*, all precondition axioms are of the simplified form

$$Poss(A(\vec{x}), s) \equiv \pi_A[s]$$

with the understanding that an action always ends in the successor situation $Do(A(\vec{x}), s)$, so that the above is a mere abbreviation of the standard axiom

$$Poss(A(\vec{x}), s, t) \equiv \pi_A[s] \wedge t = Do(A(\vec{x}), s)$$

The effects of actions are encoded by so-called *successor state axioms*, which are a compact form of general effect axioms under the assumption that all actions are deterministic. A single successor state axiom always defines the truth-value of a specific fluent in a new situation relative to both the preceding situation and the action that has been performed. Formally, if F is a fluent function, then a successor state axiom for this fluent is a formula

$$Holds(F(\vec{y}), Do(a, s)) \equiv \gamma_F^+[a, s, \vec{y}] \vee (Holds(F(\vec{y}), s) \wedge \neg\gamma_F^-[a, s, \vec{y}]) \qquad (3.11)$$

Here, γ_F^+ is a state formula describing the conditions on situation s, action a, and parameters $\vec{y}$ under which $F(\vec{y})$ is a positive effect of the action. Likewise, γ_F^- describes the conditions under which the fluent is a negative effect. Although these axioms appear considerably different from the effect axioms we have considered so far, it is possible to rewrite a set of successor state axioms in a way that makes clear that they are a specific instance of a set of general effect axioms. Let, to this end, $A(\vec{x})$ be an action, then the successor state axioms (3.11) together are mapped onto this general effect axiom:

$$Poss(A(\vec{x}), s, t) \supset (\forall f)\, [\bigvee_F (\exists \vec{y})\, (f = F(\vec{y}) \wedge \Gamma_{A,F}[\vec{x}, s]) \supset Holds(f, t)$$
$$\wedge$$
$$(\forall f)\, [\bigvee_F (\exists \vec{y})\, (f = F(\vec{y}) \wedge \neg\Gamma_{A,F}[\vec{x}, s]) \supset \neg Holds(f, t)$$

(3.12)

Here, the disjunctions $\bigvee_F$ range over all fluent functions F and $\Gamma_{A,F}[\vec{x}, s]$ stands for the formula

$$\gamma_F^+[a/A(\vec{x}), s, \vec{y}] \vee (Holds(F(\vec{y}), s) \wedge \neg\gamma_F^-[a/A(\vec{x}), s, \vec{y}])$$

Intuitively, $\Gamma_{A,F}$ is true if and only if, according to its successor state axiom, fluent F holds after action A. In this way, axiom (3.12) summarizes all positive and negative effects of $A(\vec{x})$ by instantiating the successor state axioms for each fluent by this action.

A domain axiomatization in the Situation Calculus is accompanied by foundational axioms which formally define the underlying time structure:

$$(\forall s)\, S_0 \leq s$$
$$(\forall a, a', s, s')\, (Do(a, s) = Do(a', s') \supset a = a' \wedge s = s')$$
$$(\forall a, s, s')\, (s < Do(a, s') \equiv s \leq s')$$

(3.13)

These axioms characterize a branching, tree-like time structure rooted in S_0 and where the partial ordering $s < t$ indicates that t can be reached from s by further actions.

Precondition and successor state axioms can be straightforwardly encoded as logic programs. As an example, the following clauses form a suitable logic program for the background theory in the mail delivery world:

```
poss(go(up),S)    :- holds(at(R),S), R<6.
poss(go(down),S)  :- holds(at(R),S), R>1.
poss(pick(P,B),S) :- holds(at(R),S), holds(request(P,R,R1),S),
                     holds(empty(B),S).
poss(drop(B),S)   :- holds(at(R),S), holds(carries(B,P,R),S).

holds(at(R),do(A,S)) :- holds(at(R1),S), R=R1+1, A=go(up)
holds(at(R),do(A,S)) :- holds(at(R1),S), R=R1-1, A=go(down)
holds(at(R),do(A,S)) :- holds(at(R),S), not A=go(D).
```

```
holds(empty(B),do(A,S)) :- A=drop(B)
holds(empty(B),do(A,S)) :- holds(empty(B),S), not A=pick(P,B).


holds(carries(B,P,R),do(A,S))   :- A=pick(P,B), holds(request(P,R1,R),S)
holds(carries(B,P,R),do(A,S))   :- holds(carries(B,P,R),S), not A=drop(B).


holds(request(P,R1,R2),do(A,S)) :- holds(request(P,R1,R2),S),
                                    not A=pick(P,B).
```

The reader may verify that with regard to the standard completion semantics of these clauses (cf. Section 2.2), a comparison of the successor state axioms via scheme (3.12) shows that the encoding is equivalent to the precondition and effect axioms we used in Section 3.3.1. The background theory is completed by the encoding of a particular initial situation, e.g.

```
holds(at(1),s0).
holds(empty(b1),s0).
holds(empty(b2),s0).
holds(empty(b3),s0).
holds(request(p1,1,2),s0).
...
holds(request(p9,6,4),s0).
```

The resulting logic program allows us to derive executability of actions and state properties in any concrete situation, e.g.

```
poss(pick(p1,b3),s0),
holds(carries(b3,p1,2),do(go(up),do(pick(p1,b3),s0)))
```

The computation method implicit in the definition of successor state axioms is known as *regression*: a query of the form $Holds(f, Do(\alpha_k, Do(\alpha_{k-1}, \ldots, Do(\alpha_1, S_0) \ldots)))$ is derived by repeatedly applying successor state axioms, by which the situation term is successively reduced, first to $Holds(f, Do(\alpha_{k-1}, \ldots, Do(\alpha_1, S_0) \ldots))$ and then all the way down to S_0, for which all state properties can be decided by the given initial state.

   Based on a logic program for computing the executability of actions and evaluating conditionals in GOLOG programs, a logic program that acts as an interpreter can be obtained by a direct translation of the semantics for the individual programming constructs; see Figure 3.5. The various clauses of this generic interpreter are straightforward encodings of the semantic definition of the various GOLOG programming constructs given in Section 3.4.

```
do([],S,S).
do(A,S,do(A,S)) :- poss(A,S).

do([E|L],S,S1) :- do(E,S,S2), do(L,S2,S1).

do(?(P),S,S) :- holds(P,S).

do(E1#E2,S,S1) :- do(E1,S,S1).
do(E1#E2,S,S1) :- do(E2,S,S1).

do(pi(V,E),S,S1) :- sub(V,X,E,E1), do(E1,S,S1).

do(star(E),S,S1) :- do([]#[E,star(E)],S,S1).

do(if(P,E1,E2),S,S1) :- do([?(P),E1]#[?(neg(P)),E2],S,S1).

do(while(P,E),S,S1) :- do([star([?(P),E]),?(neg(P))],S,S1).

do(P,S,S1) :- proc(P,E), do(E,S,S1).
```

FIGURE 3.5: A generic GOLOG interpreter. Sequential composition of program statements is encoded by a list using the standard Prolog list notation, [Head|Tail]. In particular, the empty program is represented by the empty list [ ]. Furthermore, the hash symbol and the keywords pi and star stand for, respectively, nondeterministic choice of sub-programs, nondeterministic choice of arguments, and nondeterministic iteration. Auxiliary predicate sub(V, X, E, E1) means that the GOLOG statement E1 is as E but with term V substituted by new variable X. It is assumed that the procedures of a GOLOG program are encoded using the predicate proc(Name, Body). For the sake of simplicity, clauses for evaluating non-atomic tests have been omitted.

3.6 EXTENSIONS: CONCURRENCY AND INTERRUPTS

3.6.1 Syntax

Basic GOLOG has a number of restrictions which make it difficult to write programs for more complex applications. Most notably, it is assumed that actions are strictly sequential and that the agent for which a program is written is the only acting entity in the environment. The language ConGOLOG (for *concurrent GOLOG*) augments the basic language by commands for dealing with concurrent executions and interrupts. The latter allows us to account for changes in the environment caused by other agents. Figure 3.6 shows the additional commands.

Example: Mail Delivery in a Dynamic Environment

The control program for the mail delivery robot in Section 3.2 has been written under the assumption that all delivery requests are given initially and that the work is done once all

Command	Meaning
$\delta_1 \parallel \delta_2$	concurrent execution
$\delta_1 \rangle\!\rangle \delta_2$	concurrent execution with priority
$\delta^{\parallel}$	concurrent iteration
$\langle \phi \rightarrow \delta \rangle$	interrupt

FIGURE 3.6: The additional syntactic elements of ConGOLOG programs. The expression ϕ is a logic formula based on the fluents of the domain, and the sub-programs δ, δ_1, δ_2 are recursively defined using all elements given here and those from basic GOLOG.

requests have been carried out. In a practical setting, however, it should be possible at any time to dynamically issue a new request (or, for that matter, to cancel an existing one). The robot must then be on the alert at any time and react sensibly to changes as soon as they occur. Most agents are in fact embedded in an environment which includes other active entities, be it humans, fellow agents, opponent players, etc. As a consequence, some state properties may not be under the sole control of one agent. Agents must therefore take into account actions besides their own when maintaining their internal world model.

From the subjective perspective of an agent, an action is *exogenous* if it is not performed by the agent itself but does affect fluents that are relevant to the agent. For a dynamic version of the control program for the mail delivery robot, we introduce two new actions for the addition and cancellation of requests, and an action for the robot to idly wait for requests if there is none at the beginning.

Symbol	Type
AddRequest	PACKAGE $\times$ ROOM $\times$ ROOM $\mapsto$ ACTION
CancelRequest	PACKAGE $\mapsto$ ACTION
Idle	$\mapsto$ ACTION

The interaction can happen at any time during the execution of the control loop. In order to account for this, the GOLOG program is extended by the following procedure:

proc *Interaction*
$\quad \pi p.\, \pi r_1.\, \pi r_2.\, AddRequest(p, r_1, r_2) \mid \pi p.\, CancelRequest(p)$
endProc;

That is to say, an interaction consists of the addition of an arbitrary request, or a cancellation. The main body is then replaced by the parallel execution of the usual control routine and any number of interactions.

proc *Deliver* ... **endProc**;

. . .

proc *Interaction* ... **endProc**;
Interaction* ∥ (**while** ¬(∃p, r_1, r_2) *Request*(p, r_1, r_2) **do** *Idle* **endWhile**; *Control*)

In this way, the interaction part runs asynchronously with, and outside of, the main control program. The program itself acts as a simulator, because it generates arbitrary, nondeterministic interactions. Of course the addition and cancellation of requests need to follow some rules, too. Specifically, a request can only be added for a package that has not already been requested or collected, and only those requests can be canceled that are actually present. Idling, on the other hand, has no preconditions. Formally,

$$
\begin{aligned}
Poss(AddRequest(p, r_1, r_2), s, t) \equiv\ & t = Do(AddRequest(p, r_1, r_2), s) \land \\
& \neg(\exists r_1', r_2')\, Holds(Request(p, r_1', r_2'), s) \land \\
& \neg(\exists b, r')\, Holds(Carries(b, p, r'), s) \land \\
& r_1 \neq r_2 \qquad\qquad (3.14) \\
Poss(CancelRequest(p), s, t) \equiv\ & t = Do(CancelRequest(p), s) \land \\
& (\exists r_1, r_2)\, Holds(Request(p, r_1, r_2), s) \\
Poss(Idle, s, t) \equiv\ & t = Do(Idle, s)
\end{aligned}
$$

Whenever an addition or cancellation happens, the internal state variables for the control program need to be updated accordingly. Idling has no effect at all (except that time progresses, of course). This is formalized in the following effect axioms for the three new actions:

$$
\begin{aligned}
&Poss(AddRequest(p, r_1, r_2), s, t) \supset \\
&\quad (\forall f)[f = Request(p, r_1, r_2) \lor Holds(f, s) \supset Holds(f, t)] \\
&\quad \land \\
&\quad (\forall f)[f \neq Request(p, r_1, r_2) \land \neg Holds(f, s) \supset \neg Holds(f, t)] \\
&Poss(CancelRequest(p), s, t) \supset \\
&\quad (\exists r_1, r_2)\,(Holds(Request(p, r_1, r_2), s) \land \\
&\qquad (\forall f)[Holds(f, s) \land f \neq Request(p, r_1, r_2) \supset Holds(f, t)] \qquad (3.15) \\
&\qquad \land \\
&\qquad (\forall f)[f = Request(p, r_1, r_2) \lor \neg Holds(f, s) \supset \neg Holds(f, t)]) \\
&Poss(Idle, s, t) \supset \\
&\quad (\forall f)[Holds(f, s) \supset Holds(f, t)] \\
&\quad \land \\
&\quad (\forall f)[\neg Holds(f, s) \supset \neg Holds(f, t)]
\end{aligned}
$$

3.6.2   Semantics

The semantics of basic GOLOG is based on the definition of programs as macros, which are unfolded into a single formula in the underlying action calculus. The introduction of concurrency and interrupts requires a different form of semantics, which is generally known as *transitional semantics*. It is based on the definition of a single execution step of a program. Formally, the semantics is given by a predicate $Trans(\delta, s, \delta', s')$. The intuitive meaning is that, for a program δ and a situation s, executing one step of δ takes one from s to situation s', and program δ' is what remains of program δ after this one step. This definition is accompanied by a predicate $Final(\delta, s)$, with the intended meaning that δ can be considered completed in situation s.

For the sake of simplicity, we again ignore the introduction of procedures and tacitly assume that a ConGOLOG program is just a main body. The definition of the transition relation is as follows:

$$Trans(\mathbf{nil}, s, \delta', s') \equiv False \qquad \text{empty program}$$
$$Trans(a, s, \delta', s') \equiv Poss(a, s, s') \wedge \delta' = \mathbf{nil} \wedge s' = Do(a, s) \qquad \text{primitive action}$$
$$Trans(\phi?, s, \delta', s') \equiv \phi[s] \wedge \delta' = \mathbf{nil} \wedge s' = s \qquad \text{test}$$
$$Trans(\delta_1; \delta_2, s, \delta', s') \equiv (\exists \delta_1')\,(Trans(\delta_1, s, \delta_1', s') \wedge \delta' = (\delta_1'; \delta_2)) \qquad \text{sequence}$$
$$\vee$$
$$Final(\delta_1, s) \wedge Trans(\delta_2, s, \delta', s')$$
$$Trans(\delta_1 | \delta_2, s, \delta', s') \equiv Trans(\delta_1, s, \delta', s') \vee Trans(\delta_2, s, \delta', s') \qquad \text{choice 1}$$
$$Trans(\pi x.\delta(x), s, \delta', s') \equiv (\exists x)\, Trans(\delta(x), s, \delta', s') \qquad \text{choice 2}$$
$$Trans(\delta^*, s, \delta', s') \equiv (\exists \delta'')\,(Trans(\delta, s, \delta'', s') \wedge \delta' = (\delta''; \delta^*)) \qquad \text{sequential iteration}$$
$$Trans(\delta_1 \| \delta_2, s, \delta', s') \equiv (\exists \delta)\,(Trans(\delta_1, s, \delta, s') \wedge \delta' = (\delta \| \delta_2)) \qquad \text{concurrency}$$
$$\vee$$
$$(\exists \delta)\,(Trans(\delta_2, s, \delta, s') \wedge \delta' = (\delta_1 \| \delta))$$
$$Trans(\delta_1 \rangle\!\rangle \delta_2, s, \delta', s') \equiv (\exists \delta)\,(Trans(\delta_1, s, \delta, s') \wedge \delta' = (\delta \rangle\!\rangle \delta_2)) \qquad \text{priority}$$
$$\vee$$
$$(\exists \delta)\,(Trans(\delta_2, s, \delta, s') \wedge \delta' = (\delta_1 \rangle\!\rangle \delta))$$
$$\wedge \neg(\exists \delta'', s'')\, Trans(\delta_1, s, \delta'', s'')$$
$$Trans(\delta^\|, s, \delta', s') \equiv (\exists \delta'')\,(Trans(\delta, s, \delta'', s') \wedge \delta' = (\delta'' \| \delta^\|)) \qquad \text{parallel iteration}$$

In addition, the conditional and loop statements from basic GOLOG are redefined in such a way that test and continuation are synchronized, in order to prevent interrupts in between:

$$Trans(\mathbf{if}\,\phi\,\mathbf{then}\,\delta_1\,\mathbf{else}\,\delta_2\,\mathbf{endIf}, s, \delta', s') \equiv \phi[s] \wedge Trans(\delta_1, s, \delta', s')$$
$$\vee$$
$$\neg\phi[s] \wedge Trans(\delta_2, s, \delta', s')$$
$$Trans(\mathbf{while}\,\phi\,\mathbf{do}\,\delta\,\mathbf{endWhile}, s, \delta', s') \equiv (\exists \delta'')\,(\phi[s] \wedge Trans(\delta, s, \delta'', s') \wedge$$
$$\delta' = (\delta''; \mathbf{while}\,\phi\,\mathbf{do}\,\delta\,\mathbf{endWhile}))$$

In contrast to macro definition (3.10), where test and continuation are separated by the sequential operator ";", the new definition requires that the transition continues in the same situation s to which the test is applied. This guarantees that no concurrent process or interrupt can invalidate the test condition prior to the execution of the body of an if-statement or a while-statement. Finally, interrupts can be defined using the other constructs as follows:

$$\langle \phi \rightarrow \delta \rangle \stackrel{\text{def}}{=} \textbf{while } \textit{True } \textbf{do}$$
$$\textbf{if } \phi \textbf{ then } \delta \textbf{ else } \textit{False?} \textbf{ endIf}$$
$$\textbf{endWhile}$$

By this definition, an interrupt is blocked as long as condition ϕ is false. As soon as this condition is triggered, the interrupt repeatedly executes sub-program δ. Since both the while- and the if-statement have been defined as synchronous, the interrupt is executed immediately after the triggering condition becomes true. If it is desired that the interrupt is terminated after it has been executed once, rather than being executed repeatedly, this can be achieved with the help of a special fluent that is initially true and set to false by δ.

The accompanying predicate $Final(\delta, s)$ determines whether a program can be considered to be completed in a situation. It is defined as follows:

$$
\begin{array}{ll}
Final(\textbf{nil}, s) \equiv \textit{True} & \text{empty program} \\
Final(a, s) \equiv \textit{False} & \text{primitive action} \\
Final(\phi?, s) \equiv \textit{False} & \text{test} \\
Final(\delta_1; \delta_2, s) \equiv Final(\delta_1, s) \wedge Final(\delta_2, s) & \text{sequence} \\
Final(\delta_1 | \delta_2, s) \equiv Final(\delta_1, s) \vee Final(\delta_2, s) & \text{choice 1} \\
Final(\pi\, x.\delta(x), s) \equiv (\exists x)\, Final(\delta(x), s) & \text{choice 2} \\
Final(\delta^*, s) \equiv \textit{True} & \text{sequential iteration} \\
Final(\delta_1 \| \delta_2, s) \equiv Final(\delta_1, s) \wedge Final(\delta_2, s) & \text{concurrency} \\
Final(\delta_1 \rangle\!\rangle \delta_2, s) \equiv Final(\delta_1, s) \wedge Final(\delta_2, s) & \text{priority} \\
Final(\delta^{\|}, s) \equiv \textit{True} & \text{parallel iteration} \\
Final(\textbf{if } \phi \textbf{ then } \delta_1 \textbf{ else } \delta_2 \textbf{ endIf}, s) \equiv \phi[s] \wedge Final(\delta_1, s) & \text{synchronized if} \\
\qquad\qquad\qquad\qquad\quad \vee & \\
\qquad \neg\phi[s] \wedge Final(\delta_2, s) & \\
Final(\textbf{while } \phi \textbf{ do } \delta \textbf{ endWhile}, s) \equiv \neg\phi[s] \vee Final(\delta, s) & \text{synchronized while}
\end{array}
$$

As an example, if we start with an initial situation without any delivery requests, then our control program for the mail delivery robot can provably evolve into the following final situation:

$$Do(Drop(B_1), Do(Go(Up), Do(Pick(P_2, B_1), Do(Go(Down),$$
$$Do(CancelRequest(P_1), Do(AddRequest(P_2, 1, 2),$$
$$Do(Go(Up), Do(AddRequest(P_1, 2, 1), Do(Idle, S_0)))))))))$$

As usual, the situation has to be read in reverse order. Hence, after idling until the first request comes in, the robot heads toward office 2, from where the request has been issued. However, this request is canceled before the robot has a chance to pick it up. Therefore, the robot immediately goes back to office 1 to fetch package P_2, which has been requested in the meantime. The program ends with a successful delivery in room 2.

The definition of the semantics is completed by redefining predicate DO as the transitive closure of program transition. This requires to appeal to second-order logic in much the same way as the definition of iteration did. Let, to this end,

$$Trans^*(\delta, s, \delta', s') \overset{\text{def}}{=} (\forall P)\,(\,[(\forall \delta_1, s_1)\, P(\delta_1, s_1, \delta_1, s_1)$$
$$\land (\forall \delta_1, \ldots, s_3)\,(P(\delta_1, s_1, \delta_2, s_2) \land Trans(\delta_2, s_2, \delta_3, s_3)$$
$$\supset P(\delta_1, s_1, \delta_3, s_3))]$$
$$\supset P(\delta, s, \delta', s'))$$

Then

$$DO(\delta, s, s') \overset{\text{def}}{=} (\exists \delta')\,(Trans^*(\delta, s, \delta', s') \land Final(\delta', s')).$$

3.7   ACTION PROGRAMS WITH SENSING

The programs we have considered thus far were all written under the tacit assumption that the agent has complete knowledge of all relevant features of its environment. Even in the setting where users may add or cancel requests, the execution of the control program requires that the agent knows about the occurrence of these actions, so that its internal world model is always in accordance with the actual world. Many realistic environments do not comply with the ideal case of complete knowledge. In highly dynamic environments, for instance, changes may happen without the agent always being aware of them. Competitive environments, too, are often characterized by an information asymmetry among the participants. The distinction between complete and incomplete information is also made in game playing, where for example most card games are characterized by the fact that the players have only partial information about the distribution of the cards.

Agents with incomplete information use sensors to acquire additional information about the environment. These *sensing* actions can be used in action programs in the same way normal actions are employed, except that the former normally do not affect the environment but enhance the knowledge of the agent.

Example: Robot in a Maze

Suppose a robot has to find a path through an unknown maze. Initially, it has no knowledge at all of the structure of the maze, but it is equipped with a sensor that tells it whether it is facing a wall. The use of this sensor obviously does not change the position of the robot, nor does it have any effect on the environment. Nonetheless, without a sensor the problem could not be solved, because the robot lacks sufficient knowledge at the beginning, and sensing helps it to acquire this knowledge.

For GOLOG programs with sensing actions, the pure offline execution is no longer meaningful. The reason is that the outcome of a sensing action cannot be predicted in advance; if it were otherwise, the sensing action would not be needed. Programs with sensing actions are therefore executed online. A simple strategy to negotiate an unknown maze, for example, is to turn right whenever possible.[3] This is implemented by the following basic GOLOG program:

```
proc TakeNextStep
    TurnRight;
    SenseBlocked;
    while Blocked do
        TurnLeft;  SenseBlocked
    endWhile;
    GoForward
endProc;
```

```
while ¬At(Exit) do TakeNextStep endWhile
```

Here, *TurnRight*, *TurnLeft*, and *GoForward* are physical actions of the agent with the obvious meaning. The action *SenseBlocked* is a pure sensing action by which is checked whether the path is blocked in the direction the robot currently faces. Fluent *Blocked* shall be true whenever this is the case. Without full knowledge of the layout of the maze, this GOLOG program cannot be executed offline. Specifically, without sufficient knowledge of the truth-value of fluent *Blocked* in the relevant situations the semantics does not entail a sequence of actions which determines a successful run of the program. Online execution, on the other hand, should provide the control program with sufficient information about this fluent at the time it is needed to evaluate the condition in the loop of procedure *TakeNextStep*.

In more complex environments, however, it may be desirable to retain the advantage of offline execution. For this purpose, a *search operator* can be used, denoted by the symbol Σ. This operator allows us to execute designated parts of a program offline, having the agent search for an appropriate action sequence in specific situations prior to actually executing it. This results in an interleaved online/offline execution model. To illustrate this, consider the

[3]If the maze allows closed cycles, additional measures need to be taken to prevent the agent from walking in circles.

following GOLOG sequence:

$$\Sigma \{ (PlanA \mid PlanB); \ SubGoal?\}; \ (ActC \mid ActD); \ Test?$$

This is to be understood as follows. First, the control program nondeterministically executes either *PlanA* or *PlanB* in such a way as to make sure that *SubGoal* is true afterwards. Being in the scope of the search operator, this nondeterministic choice is made offline, which allows us to select the right plan to achieve this goal, if there is any. After leaving the scope of the search operator, the chosen plan is actually executed according to the online-execution paradigm. This is followed by a nondeterministic choice between *ActC* and *ActD*, but now the selection has to be made online. That is to say, the control program commits to its choice prior to verifying that *Test* holds. In practice, this requires some care on the side of the programmer, who needs to make sure that the program terminates no matter which nondeterministic choice is made outside of a search operator.

The effects of sensing actions are axiomatized with the help of a special predicate. For the sake of simplicity, we only consider binary sensing actions, that is, which tell the agent whether a specific state property is true or false. The effects of these actions are described via the predicate $SF(a, s)$, which, if true for situation s, indicates that (sensing) action a would result in the answer "yes." For the sake of uniformity, SF may range over all actions, whether they involve sensing or not, and then be defined as "true" in all situations for every non-sensing action of a domain. For instance, the background axiomatization for the robot-in-a-maze domain may include the formulas

$$SF(SenseBlocked, s) \equiv Holds(Blocked, s)$$

and $SF(a, s) \equiv True$ for all other actions a.

The definition of the execution of a GOLOG program with sensing is based on the concept of a *history*. This is a sequence of pairs (a, v), where a is an action and v a sensing result, that is, either of the values *true* or *false*. Intuitively, a history like

$$(TurnRight, true), (SenseBlocked, false), (GoForward, true), \ldots \qquad (3.16)$$

describes a sequence of actions along with an actual sensing result associated with each action. We use two abbreviations related to histories: if h is a history

$$(a_1, v_1), (a_2, v_2), \ldots, (a_n, v_n)$$

then by $End[h, s]$ we mean the situation $Do(a_n, \ldots, Do(a_2, Do(a_1, s))\ldots)$. By $Sensed[h, s]$ we denote the formula that encodes the entire sensing information given in the history, that is,

$$\{\neg\}SF(a_1, Do(a_1, s)) \wedge \{\neg\}SF(a_2, Do(a_2, Do(a_1, s))) \wedge \ldots \wedge \{\neg\}SF(a_n, End[h, s])$$

where the "$\neg$" is placed just in case v_i is *false*. Thus, for instance, the history in (3.16) provides the controller with the following sensing information:

$$Turn \wedge$$
$$\neg Holds(Blocked, Do(SenseBlocked, Do(TurnRight, S_0))) \wedge$$
$$Turn \wedge \ldots$$

Based on the concept of a history, the semantics for GOLOG is easily adapted to programs with sensing actions as follows. Recall the expression $Trans(\delta, s, \delta', s')$, which means that, given a program δ and a situation s, executing one step of δ takes one from s to situation s', and program δ' is what remains of the program δ after this one step. In the presence of sensing, the validity of a transition step may depend on the results of the preceding sensing actions, that is, on the history. Specifically, given a history h starting in some situation s, a transition step is possible if the domain axiomatization along with $Sensed[h, s]$ entails that $Trans(\delta, End[h, s], \delta', s')$ for some δ' and s'. Similarly, a program may terminate after a history h if the domain axiomatization and $Sensed[h, s]$ entail $Final(\delta, End[h, s])$. The only actual extension of the semantics is required for the search operator, whose semantics is defined as follows:

$$Trans(\Sigma\delta, s, \delta', s') \equiv (\exists\delta'')(Trans(\delta, s, \delta'', s') \wedge \delta' = \Sigma\delta'' \wedge$$
$$(\exists\delta_e, s_e)(Trans^*(\delta'', s', \delta_e, s_e) \wedge Final(\delta_e, s_e)))$$
$$Final(\Sigma\delta, s) \equiv Final(\delta, s)$$

Put in words, a transition is possible under the search operator if (δ, s) can evolve to (δ'', s') *and* it is possible to reach some final configuration (δ_e, s_e) from there. The latter condition is precisely what characterizes the search operator: prior to making a single transition step, it needs to be ensured that the sub-program which is subject to search can be successfully completed.

3.8   EXERCISES

3.1. Consider the precondition and effect axioms for the mail delivery world along with the *action closure axiom*

$$(\forall a)[(\exists d) a = Go(d) \vee (\exists p, b) a = Pick(p, b) \vee (\exists b) a = Drop(b)]$$

Suppose that the robot is equipped with three mail bags, which are all empty initially, and that there are just two initial requests. Use the *induction* axiom

$$(\forall P)\,\{\,P(S_0) \wedge (\forall a, s)\,[P(s) \supset P(Do(a, s))] \supset (\forall s)\,P(s)\,\}$$

to formally prove that in this case there is no reachable situation without an empty mail bag!

3.2. (a) Let $\delta_{mailbot}$ denote the GOLOG program for the mail delivery robot from Section 3.6, which accounts for the dynamic addition and cancellation of requests. Find an initial situation and an infinite sequence of situations $S_1, S_2, S_3, \ldots$ such that $Trans(\delta_{mailbot}, S_0, \delta_i, S_i)$ holds for all $i = 1, 2, 3, \ldots$ (and the appropriate $\delta_1, \delta_2, \delta_3, \ldots$) and where one of the initial requests is never picked up!

(b) A general solution to this problem is to give requests increasing priority the longer they have been around. One way of implementing this is to extend the domain axiomatization by a fluent $Time(t)$ such that t indicates the number of Go actions the robot has performed since the beginning. Redefine, to this end, the fluents $Request(p, r_1, r_2)$ and $Carries(b, p, r)$ so as to include the information at what time the request in question has been issued, and modify the effect axioms accordingly! Program a refined control strategy by which every request is guaranteed to be carried out eventually! Take care also that the robot does not carry around some package forever!

3.3. Specify the rules of the game Tic-Tac-Toe by appropriate precondition and effect axioms using the relational fluents

Symbol	Type
Cell	$\{1, 2, 3\} \times \{1, 2, 3\} \times \{X, O\} \mapsto$ FLUENT
Control	$\{XPlayer, OPlayer\} \mapsto$ FLUENT

and the actions

Symbol	Type
Mark	$\{X, O\} \times \{1, 2, 3\} \times \{1, 2, 3\}$

Modeling the moves of the opponent by exogenous actions, write a ConGOLOG for optimal play!

3.4. Specify the rules of the game Connect Four (cf. Figure 3.7) in the Situation Calculus using *functional* fluents! Define, to this end, a variant of the successor state axioms (3.11)

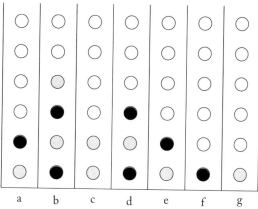

FIGURE 3.7: The two-player game Connect Four: Red and Black take turn in dropping discs of their own color into one of the seven columns. The player wins who is the first to have four connecting discs horizontally, vertically, or diagonally. In the present situation, for example, Black should drop a disc in column c to prevent Red from completing a diagonal b–e. Incidentally, this happens to be a forced win for Black, because Red cannot hinder it to get four discs in a row with the next move.

suitable for functional fluents! Accordingly, take the GOLOG interpreter from Section 3.5 and replace the basic predicate holds(F, S) by val(F, V, S) for functional fluents, meaning that fluent F has value V in situation S. Encode the precondition and successor state axioms accordingly and write and implement a GOLOG program, possibly including the search operator, that plays Connect Four well—ideally so well as to win whenever it has the first move!

CHAPTER 4

Action Programs and Planning

Planning is the process of searching for a suitable, goal-oriented strategy by the agent itself. An agent that plans entertains the effects of various possible action sequences before starting to act. The ability to find plans on their own increases the autonomy of agents and makes programming much easier in cases where it would require considerable effort to implement a pre-defined strategy which is good for all situations that the agent may encounter.

4.1 PLANNING WITH PLAN SKELETONS

We consider the basic definition of a planning problem, where the planning agent has complete control of the part of its environment that is relevant for the planning task at hand. In particular, the agent plans under the assumption that no other agent may hinder it to take a suitable course of actions. Single-player games provide a good metaphor for this kind of planning problems. Consider, as an example, the puzzle depicted in Figure 4.1. This game can be formulated as a search problem in an action domain as follows. The various positions (states) in this game are described with the help of a single fluent, $Cell(x, v)$, which means that at position $x = a, \ldots, h$ lie $v \in \{0, 1, 2\}$ coins. The initial state is then given by the formula

$$Holds(f, S_0) \equiv f = Cell(a, 1) \vee f = Cell(b, 1) \vee f = Cell(c, 1) \vee f = Cell(d, 1) \vee$$
$$f = Cell(e, 1) \vee f = Cell(f, 1) \vee f = Cell(g, 1) \vee f = Cell(h, 1)$$

There is only one action the agent can perform, $Jump(x, y)$, meaning to take the coin on x and place it on y. Using situations as the underlying time structure, these are the preconditions of this action according to the rules.

$$Poss(Jump(x, y), s, t) \equiv t = Do(Jump(x, y), s) \wedge$$
$$Holds(Cell(x, 1), s) \wedge Holds(Cell(y, 1), s) \wedge$$
$$CoinsBetween(x, y, s) = 2$$

Auxiliary function $CoinsBetween(x, y, s)$ is assumed to count the number of coins that lie between the positions x and y in situation s. The effect of jumping is to change the number of

$$a \quad b \quad c \quad d \quad e \quad f \quad g \quad h$$

FIGURE 4.1: A simple one-player game: starting with eight coins in a row, jump with a single coin over two coins onto another single coin. Repeat until you end up, after four moves, with four stacks of two coins each. For example, it is a valid first move to place the coin from a onto the one in d, and then the coin from c can be put onto the one in e. This, however, would lead to a position in which no further move can be taken.

coins at the two positions as follows:

$$Poss(Jump(x, y), s, t) \supset$$
$$(\forall f)[f = Cell(x, 0) \vee f = Cell(y, 2) \vee (Holds(f, s) \wedge f \neq Cell(x, 1) \wedge f \neq Cell(y, 1))$$
$$\supset Holds(f, t)]$$
$$\wedge$$
$$(\forall f)[f = Cell(x, 1) \vee f = Cell(y, 1) \vee (\neg Holds(f, s) \wedge f \neq Cell(x, 0) \wedge f \neq Cell(y, 2))$$
$$\supset \neg Holds(f, t)]$$

Put in words, the number of coins in x changes from 1 to 0 and the number of coins in y from 1 to 2. The goal is to reach a state in which all stacks are of size two, that is,

$$Goal(s) \stackrel{\text{def}}{=} (\forall x, v)(Holds(Cell(x, v), s) \supset v = 0 \vee v = 2)$$

We have now defined all components of a planning problem: given an initial state and a definition of the possible actions, the task is to find an action sequence—a.k.a. plan—that leads to a situation in which the goal is satisfied.

This example planning problem can actually be easily solved by a complete search through the entire tree of situations. The branching factor is comparably low (8 at the beginning and then decreasing) and the reachable situations have a maximal length of 4. The following simple GOLOG program can thus be used to have the agent solve the puzzle on its own:

$$\textbf{while } \neg Goal \textbf{ do}$$
$$\pi a \, . \, a$$
$$\textbf{endWhile}$$

If executed offline (in ConGOLOG this sequence would have to be included in the scope of the search operator Σ; cf. Section 3.7), any successful run of this nondeterministic program establishes a sequence of executable actions that leads to a situation in which the goal is satisfied. In other words, this generic program generates plans for the planning problem which is given by the definition of the goal predicate along with the appropriate background theory.

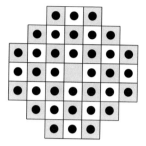

FIGURE 4.2: An instance of a non-trivial planning problem, known as Peg Solitaire: jump with one peg over another into an empty spot. Repeat until you end up with one single peg in the center square.

In contrast to this simple example, many planning problems are too complex to be tackled by blind search. Figure 4.2 shows another single-player game that falls into this category. Although this is a comparably small board, a successful plan has to be of length 35 because the goal is to get rid of all but one of the 36 pegs and every action removes one of them. Furthermore, although initially there are just 4 possible actions, the branching factor quickly increases after a few jumps as more holes become available. This results in a total search space too large to be solved in reasonable time by simple brute-force search.

There are several well-known, domain-independent approaches to tackle large planning problems by reducing the search space. An example is the use of hash tables, also known as transposition tables, in which all intermediate states are saved. This allows us to cut off branches in the search tree which lead to states that have been searched earlier via a different action sequence. A more involved approach is to use symmetries to avoid searching twice different but symmetric action sequences. This requires to find suitable symmetries in a problem, like the obvious rotational and reflectional symmetry on the Solitaire board. Since symmetries can often be derived fully automatically from the formal specification of a planning problem, this search strategy also falls into the category of domain-independent techniques.

But while domain-independent planning methods may help in some cases, they often do not scale up well to larger problem instances and thus do not suffice to solve problems that occur in practice. A much better performance can be obtained if the type of planning problem is known in advance so that strong, tailor-made search strategies can be integrated into the planning algorithm.

Example: A Planning Strategy for Peg Solitaire

A good, domain-dependent planning strategy for any kind of Peg Solitaire board is to successively identify and solve certain small patterns. Figure 4.3(a) depicts three common patterns, which suffice to solve most Peg Solitaire boards. A search strategy based on these patterns is obviously domain-dependent, but on the other hand it is still general enough to be applicable to

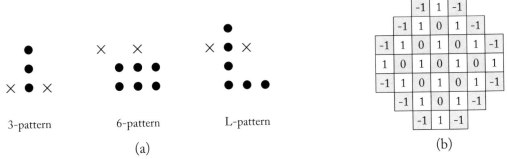

3-pattern   6-pattern   L-pattern

(a)   (b)

FIGURE 4.3: (a) Each of these three patterns for Peg Solitaire can be completely removed by a sequence of jumps given that one of the cells marked by "×" contains a peg. (b) A weight function for the Peg Solitaire board, which has the property that positions in which the weighted sum of all occupied squares falls below zero lead to a dead end.

any type of board and initial configuration. The following GOLOG program has the planning agent systematically locate and clear patterns on an arbitrary Peg Solitaire board until a goal position is reached:

> **proc** *SolvePattern*(p)
>   **while** $\neg Solved(p)$ **do**
> $\pi x.\, \pi y.\, \pi z.\, (\, y \in p\, ?;\; Jump(x, y, z)\,)$
>   **endWhile**
> **endProc**;
>
> **while** $\neg Goal$ **do**
> $\pi p.\, (IsPattern(p)\, ?;\; SolvePattern(p)\,)$
> **endWhile**

In this program, $Jump(x, y, z)$ represents the action of jumping with the peg currently in position x over y into square z. Expression $IsPattern(p)$ is an auxiliary fluent that is assumed to be true if region p matches against one of the given patterns; $Solved(p)$ should be defined in such a way as to hold if pattern p has been completely removed from the board; and $x \in p$, finally, denotes that cell x is contained in the set of cells p describing a pattern. These expressions need to be formulated as fluents in the underlying action theory because GOLOG does not include language elements that allow us to define auxiliary predicates outside of the underlying domain language.[1]

---

[1] In Chapter 5 on declarative actions programs, we show how this restriction is overcome by combining reasoning about actions with general logic programming.

The GOLOG program describes a sophisticated divide-and-conquer solution to a complex planning problem. It is obviously nondeterministic, hence defines a tree of possible runs when executed offline—which is necessary since not all sequences of pattern removals allow us to clear the entire board. On the other hand, it is much more focused than the generic search program considered earlier, which includes just any executable action sequence. The new, tailor-made GOLOG program can be viewed as a compact way of defining a *plan skeleton*. Thanks to its nondeterministic programming constructs, GOLOG provides an elegant way to program such skeletons for a given class of planning problems. It should be noted that the strategy of using patterns is not *strongly complete*, meaning that there are successful action sequences which lie outside the defined skeleton. This is so because there may be useful patterns other than the ones defined for the program and also because in general the process of clearing an individual pattern may be interleaved with the solution of another pattern, whereas the control program requires to completely remove one pattern before turning to the next.

Another way to cut down the search space for a planning problem is to use a generic definition of a plan skeleton which defines local constraints on the selection of an action. This can be illustrated with an alternative solution to Peg Solitaire. Figure 4.3(b) shows what is known as "resource count." This function has the property that it can only decrease after a legal move in any situation. Hence, given that the resource count in the final position is 0, there can be no solution in which the sum of the values of all occupied cells turns negative. This gives rise to a local cut-off criterion, namely, as soon as an action leads to a position that determines a negative resource count. Constraints of this nature can be combined with the following generic GOLOG program for planning:

$$\textbf{while } \neg Goal \textbf{ do}$$
$$\pi\, a\,.\, a;\, Allowed?$$
$$\textbf{endWhile}$$

In the Peg Solitaire case this program would be accompanied by the definition

$$Holds(Allowed, s) \equiv ResourceCount(s) \geq 0$$

where $ResourceCount(s)$ is the sum of the weights of all occupied positions in situation s. Since there can be no solution to Solitaire in the course of which the resource count becomes negative, this plan skeleton encompasses all successful plans. Hence, as opposed to the strategy based on patterns, this alternative heuristics is strongly complete.

4.2 PLANNING WITH PRUNING

The last, generic GOLOG program for planning uses constraints on situations as a domain-dependent pruning technique, by which unsuccessful branches of the search tree are cut off at an early stage. These constraints are *local* because they refer solely to the current situation. Complex planning problems often require a more expressive pruning technique, using constraints that combine properties of past actions with the present situation. *Linear Temporal Logic* provides a purely declarative way of encoding such search control rules.

Linear Temporal Logic extends classical logic by the following *modal* operators:

symbol	meaning
$\bigcirc \phi$	ϕ holds next
$\square \phi$	ϕ always holds
$\Diamond \phi$	ϕ eventually holds
$\phi \, \mathcal{U} \, \psi$	ϕ holds until ψ

The following formula, for example, may express the control rule that whenever some container x is in a vehicle y, then the container should stay there until the vehicle reaches the goal location of x:

$$\square \, [\, In(x, y) \supset In(x, y) \, \mathcal{U} \, At(y, GoalLocation(x)) \,]$$

Linear, temporal formulas are interpreted w.r.t. infinite sequences $S_0, S_1, S_2, \ldots$ of situations as follows. A formula ϕ is *true* in a particular situation S_i, written $S_i \models \phi$, if

- $Holds(\phi, S_i)$, where ϕ is atomic;
- $S_{i+1} \models \psi$, where ϕ is $\bigcirc \psi$;
- $S_j \models \psi$ for all $j \geq i$, where ϕ is $\square \psi$;
- $S_j \models \psi$ for some $j \geq i$, where ϕ is $\Diamond \psi$;
- $S_j \models \psi_2$ for some $j \geq i$ such that $S_k \models \psi_1$ for all $i \leq k < j$, where ϕ is $\psi_1 \, \mathcal{U} \, \psi_2$;
- $S_i \models \psi$, where ϕ is $\neg \psi$; likewise for the other logical connectives.

Example: A Logistics Problem

To illustrate the usefulness of control rules, consider a problem in logistics where two types of vehicles are used to transport goods: trucks, which operate within a city, and airplanes, which operate between cities. A planning problem in this setting consists of an initial configuration of a set of goods at various locations and the goal to redistribute these objects to their individual destination. The actions in this domain are to load and unload objects into and off vehicles, and to move trucks and fly planes to new locations:

Symbol	Type
Load	OBJECT × VEHICLE ↦ ACTION
Unload	OBJECT × VEHICLE ↦ ACTION
Move	VEHICLE × LOCATION ↦ ACTION
Fly	VEHICLE × LOCATION ↦ ACTION

For the sake of simplicity, assume that trucks and planes have sufficient capacity, so that resource constraints need not be considered. Still, due to exponential explosion, a complete tree search in this domain is practically impossible. Even a small problem with, say, just ten goods, four cities, in each city two locations and a truck, and three airplanes has a total state space of 6×10^{14} and a search tree that is larger by several magnitudes.

A few simple control rules, however, suffice to make this planning problem manageable. When searching for a plan, the agent should consider to

1. move a vehicle only if there are no objects at its present location which need to be loaded into or unloaded from the vehicle;

2. move a vehicle only to locations where it needs to pick up or to drop goods;

3. load objects only into the kind of vehicle (truck or plane) with which they need to be transported;

4. unload objects only where they need to be unloaded.

The formalization of these rules with the help of temporal modalities requires two auxiliary predicates that are derived from the goal specification: $MustBeMovedBy(o, v)$ means that, according to the planning goal, object o has to be moved by the type of vehicle (truck or airplane) to which v belongs.[2] $MustBeUnloadedAt(o, l)$ means that, according to the goal, object o must be unloaded at location l.[3] With this, the control rules translate into the following temporal logic formulas, where free variables are universally quantified as usual.

---

[2]This holds for all trucks and an object whose goal location differs from its initial location, and for all airplanes and an object whose goal location is in a different city.

[3]This holds for an object and its goal location, as well as for an object and the city airport of both its initial and final location provided the object requires air transportation.

1. Blocking condition for moving a truck or flying an airplane:

$$\Box\,[\,Location(v, l) \wedge (\exists o)\,(At(o, l) \wedge MustBeMovedBy(o, v)) \supset \bigcirc Location(v, l)\,]$$
$$\Box\,[\,Location(v, l) \wedge (\exists o)\,(In(o, v) \wedge MustBeUnloadedAt(o, l)) \supset \bigcirc Location(v, l)\,]$$

The fluents $Location(v, l)$ and $At(o, l)$ mean that, respectively, vehicle v and object o are currently at l, and fluent $In(o, v)$ means that object o has been loaded into vehicle v.

2. Conditions for moving or flying a vehicle:

$$\Box\,[\,Location(v, l) \wedge \bigcirc Location(v, l') \wedge l \neq l' \supset$$
$$(\exists o)\,(At(o, l') \wedge MustBeMovedBy(o, v) \vee In(o, v) \wedge MustBeUnloadedAt(o, l'))\,]$$

3. Blocking condition for loading an object:

$$\neg MustBeMovedBy(o, v) \supset \Box\neg In(o, v)$$

4. Blocking condition for unloading an object:

$$\Box\,[\,In(o, v) \wedge Location(v, l) \wedge \neg MustBeUnloadedAt(o, l) \supset \bigcirc In(o, v)\,]$$

As control rules, linear temporal logic formulas are used to ensure that a given sequence of actions satisfies them. Every action sequence determines a sequence of situations, against which the validity of a control formula can be checked. Furthermore, the formulas can be used actively, that is, the search tree gets pruned as soon as a situation is reached which violates one of them. This is the case whenever no continuation of the current action sequence could eventually satisfy the formula in question.

Evaluating temporal logic formulas in every state can be costly, however, as it may require to check the entire past action sequence. As the search tree grows deeper, this can considerably slow down the computation of the pruning condition. Fortunately it is possible to use the control formulas in a local fashion by *progressing* them to each new situation. If progression fails, then this indicates that the current sequence of actions violates the formula, whereas if it succeeds, then a plan satisfies the formula provided the progressed formula will be satisfied by the future action sequence. For example, a rule $\Box\phi$ is valid just in case ϕ holds now *and* $\Box\phi$ holds in the next situation. Or, a rule $\Diamond\phi$ is valid just in case ϕ holds now *or* $\Diamond\phi$ holds in the future. The full, recursive definition of the result of progressing a formula ϕ, written $Progress(\phi)$, is as follows, where S refers to the current situation:

- *True* (respectively, *False*) if ϕ is atomic and $Holds(\phi, S)$ (respectively, $\neg Holds(\phi, S)$);
- ψ, if ϕ is $\bigcirc\psi$;
- $Progress(\psi) \wedge \Box\psi$, if ϕ is $\Box\psi$;
- $Progress(\psi) \vee \Diamond\psi$, if ϕ is $\Diamond\psi$;

- $Progress(\psi_2) \vee (Progress(\psi_1) \wedge \psi_1 \, \mathcal{U} \, \psi_2)$, if ϕ is $\psi_1 \, \mathcal{U} \, \psi_2$;
- $Progress(\psi_1) \wedge Progress(\psi_2)$, if ϕ is $\psi_1 \wedge \psi_2$; likewise for the other logical connectives;
- $\bigwedge_c Progress(\psi\{x/c\})$, if ϕ is $(\forall x) \, \psi$;
- $\bigvee_c Progress(\psi\{x/c\})$, if ϕ is $(\exists x) \, \psi$.

The last two items imply a restriction of the progression method. For if the scope of a quantifier includes a modal operator, then progression requires to instantiate the variable by all objects of the right sort. This is possible only in case finitely many such objects exist.

The correctness of the use of progression relies on the fact that, given a sequence $S_0, S_1, \ldots$ of situations, a formula ϕ is entailed at situation S_i just in case $Progress(\phi)$ (through S_i) is entailed at S_{i+1}. Control rules can thus be used for planning in an active and local manner, where search is stopped as soon as a rule can no longer be progressed through the current situation. A few strong control rules can cut down the search space considerably. Experiments with the logistics domain, for example, have shown that the four control rules from above allow us to manage problems with a total state space of 10^{60} and beyond.

4.3   PLANNING WITH PREFERENCES

A crucial feature of the classical definition of a planning problem is that all solutions are considered equal. The only requirement is that a plan leads to a state which satisfies the goal condition—how it does this, and which of potentially many different concrete goal states it achieves, is irrelevant. In many practical settings, this homogeneous treatment of a possible multitude of plans is often inappropriate. Two action sequences may be of very different quality, and if both achieve the goal, then the agent should always take the preferred one. To this end, a basic planning problem can be extended by *preference information* in addition to the mere goal specification. The task, then, is to find plans of high quality, that is, which are preferred over all other plans that also happen to achieve the goal.

One obvious way of expressing preference is to define a numeric objective function, which assigns a real number to all action sequences. While this is arguably the conceptually simplest way of handling preference information, the major disadvantage is that it can be quite difficult to fully quantify a given preference if this is only partially known and of qualitative nature. A more sophisticated and flexible way of formalizing preferences adopts the modal operators of the previous section and uses them to express preferences in terms of desirable, but not mandatory, properties of plans.

Consider, to this end, a formal language consisting of a set of fluents and actions for a concrete planning domain. Preferences are then defined using linear temporal logic based on fluents as atoms and two additional, special atomic expressions: `final(f)` and `occurs(a)`. The former is used to stipulate that in the final state fluent f holds, while the latter requires action a to occur next. This allows us to formulate basic desires such as the eventual occurrence of a certain action, e.g.,

$$\Diamond \text{occurs}(Load(Package_6, Truck_1))$$

or the wish that some property continues to hold until a given action is executed, as in

$$At(Truck_1, Station_3) \, \mathcal{U} \, \text{occurs}(Move(Truck_1, Airport))$$

Preferences are based on these desires, which are formally defined as follows:

Definition 4.3.1. *A basic desire formula can be*

- *a fluent;*
- *the atomic* $\text{final}(f)$, *where f is a fluent;*
- *the atomic* $\text{occurs}(a)$, *where a is an action;*
- *any combination of the above using the logical connectives and the modalities of linear temporal logic.* □

Two or more basic desire formulas can be brought into a specific order, which allows us to express preferences among alternatives.

Definition 4.3.2. *An atomic preference formula is of the form*

$$\varphi_0 \prec \varphi_1 \prec \ldots \prec \varphi_n$$

where $n \geq 0$ and the φ_i's are basic desire formulas. If $n = 0$, then a preference formula coincides with a basic desire formula. □

A simple example of an atomic preference formula is

$$\Diamond \text{occurs}(Load(Package_6, Truck_1)) \prec \Diamond \text{occurs}(Load(Package_6, Truck_2)) \qquad (4.1)$$

Put in words, all else being equal it is preferred to transport $Package_6$ with $Truck_1$ rather than with $Truck_2$.

Finally, atomic preference formulas can be combined into complex preferences with the help of special logical connectives.

Definition 4.3.3. *A general preference formula* can be

- *an atomic preference formula;*
- *a conditional $\varphi \Rightarrow \Phi$, where φ is a basic desire formula and Φ a general preference formula;*
- *a general conjunction $\Phi_1 \& \ldots \& \Phi_n$, where $n \geq 2$ and the Φ_i's are general preference formulas;*
- *a general disjunction $\Phi_1 | \ldots | \Phi_n$, where $n \geq 2$ and the Φ_i's are general preference formulas.*

$\square$

The intuitive meaning of a conditional is to say that some preference holds under certain conditions only. For example, the formula

$$\mathtt{final}(Location(Truck_2, Airport)) \;\Rightarrow\; \text{(4.1)}$$

expresses a preference for transporting $Package_6$ with the first truck (cf. formula (4.1)), if the second truck happens to end up at the airport.

The purpose of specifying preferences is to find plans that satisfy them as far as possible. Since each plan determines a sequence of situations $S_0, S_1, \ldots, S_n$, preference formulas can be evaluated against this sequence in much the same way as control rules are. For basic desire formulas, the temporal modalities ("next," "always," "eventually," and "until") are interpreted as usual, and the validity of the two special atoms is defined by

- $S_i \models \mathtt{final}(f)$ if $Holds(f, S_n)$, and
- $S_i \models \mathtt{occurs}(a)$ if $S_{i+1} = Do(a, S_i)$.

Based on the evaluation of all basic desire formulas, a *weight* ω can be defined which indicates to what extent a plan violates the given preferences. Intuitively, a weight of 0 characterizes an ideal plan, in which all preferences are satisfied. Let, to this end, $S_0, S_1, \ldots, S_n$ be the sequence of situations determined by a plan, then a basic desire formula φ determines the weight

$$\omega(\varphi) = \begin{cases} 0 & \text{if } S_0 \models \varphi \\ 1 & \text{otherwise.} \end{cases}$$

The weight of an atomic preference formula is defined as the lowest rank of a basic desire that is satisfied by the plan at hand. In other words, the more preferable desires that are not satisfied, the worse the weight gets. Formally,

$$\omega(\varphi_0 \prec \ldots \prec \varphi_n) = \begin{cases} \min\{i : \omega(\varphi_i) = 0\} & \text{if it exists} \\ n + 1 & \text{otherwise.} \end{cases}$$

Thus the weight remains 0 if the most preferred desire is satisfied, and it becomes $n + 1$ if none of the desires have been achieved. Finally, the weight of an arbitrary preference formula is defined recursively as follows:

$$\omega(\varphi \Rightarrow \Phi) = \begin{cases} \omega(\Phi) & \text{if } \omega(\varphi) = 0 \\ 0 & \text{otherwise} \end{cases}$$

$$\omega(\Phi_1 \& \ldots \& \Phi_n) = \sum_i \omega(\Phi_i)$$

$$\omega(\Phi_1 | \ldots | \Phi_n) = \min_i \omega(\Phi_i).$$

Given a preference specification, the association of a weight with every sequence of situations provides a straightforward measure by which plans can be compared according to the extent to which they take into account the preferences. An ideal plan leads to an overall weight of 0 for all preferences, and if we have two plans p_1 and p_2, then one is preferred over the other just in case it determines a lower overall weight. Planning with preferences thus means to find an action sequence which achieves the goal and at the same time has the lowest weight of all successful sequences.

Clearly, the incorporation of preferences can only add to the complexity of planning. In particular, the evaluation of the preferences for each and every successful plan can be very costly, as it normally requires to go through each action sequence to determine the associated weight. Fortunately it is possible to avoid evaluating the preference formulas separately for each plan. The idea is to resort to the progression principle again. A basic desire formula can be progressed in exactly the same way as a temporal control formula and with the following definition for the special atoms. If, as before, $Progress(\phi)$ denotes the progression through a situation S, then

- $Progress(\texttt{final}(f))$ is $\texttt{final}(f)$; and
- $Progress(\texttt{occurs}(a))$ is $\texttt{next}(a)$, with the auxiliary definition

$$S_i \models \texttt{next}(a) \text{ if } S_i = Do(a, s) \text{ for some } s;$$
$$Progress(\texttt{next}(a)) \text{ is } True \text{ if } S = Do(a, s) \text{ for some } s, \text{ else } False.$$

The progression of atomic preference formulas and general preference formulas is a straightforward generalization of this definition, applying progression to all basic desire formulas inside.

The correctness of progression relies on the fact that—w.r.t. a sequence $S_0, S_1, \ldots, S_n$ of situations—a preference formula Φ has the same weight in situation S_0 as the formula $Progress^n(\Phi)$ has in situation S_n. The preference formulas can thus be evaluated across different plans when searching for a plan. Moreover, a so-called best-first search strategy can be applied by ordering partial plans according to their current weights.

4.4   EXERCISES

4.1. Specify the 15 Puzzle (cf. Figure 4.4) as a planning problem in the Situation Calculus! Write a GOLOG program that defines a suitable plan skeleton to solve this puzzle for any given initial configuration!

4.2. Specify precondition and effect axioms for the Container Stacking Problem depicted in Figure 4.5! Find useful temporal logic formulas to be used as control rules for a forward chaining planner! Define, to this end, suitable auxiliary predicates that can

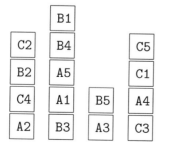

FIGURE 4.4: An instance of the 15 Puzzle: starting with the configuration shown on the left hand side, a single move consists in sliding a tile into the empty space. The problem is solved if the configuration shown on the right hand side is reached.

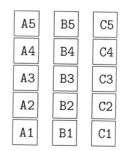

FIGURE 4.5: A Container Stacking Problem: starting with the configuration shown on the left hand side, find a minimal sequence of actions by a forklift to reach the ordered configuration shown on the right hand side. Assume that the forklift can lift only one container at a time and that there is sufficient space on the ground to put down a container.

be derived from a given goal configuration. Aim at a set of control rules which are so strong that they allow no branching at all, so that planning is *linear* in the number of containers while it is guaranteed that a plan with minimal number of actions is found!

4.3. Show the correctness of the progression operator for preference formulas: for a preference formula Φ and a sequence $S_0, S_1, \ldots, S_n$ of situations, let $w_1 = \omega(\Phi)$ w.r.t. S_0 and $\Psi = Progress^n(\Phi)$. Prove that if $w_2 = \omega(\Psi)$ w.r.t. S_n, then $w_1 = w_2$!

4.4. Extend the axiomatization of the mail delivery world by allowing users to give different priorities to their requests! Add appropriate preference formulas for priority handling! Extend the generic GOLOG interpreter from Chapter 3 by an evaluation mechanism for preference formulas, and modify the GOLOG program for the mail delivery robot so as to generate the most preferred one of all plans with minimal number of actions!

CHAPTER 5

Declarative Action Programs

5.1 AGENT LOGIC PROGRAMS

Declarative action programs combine reasoning about actions with the declarative programming paradigm. Agent Logic Programs, or short ALPs, are an amalgamation of standard logic programs with the actions and fluents of an agent and its environment. As such, an ALP looks a lot like a standard logic program, but it uses special elements which capture the dynamics of actions and change.

Similar to the procedural action programming language GOLOG, declarative action programs are based on the specification of the actions an agent can take and the fluents which describe the environment of the agent. In contrast to a procedural action programming language, however, the high-level strategy of the agent is given by a logic program that uses its own signature with arbitrary additional predicates and functions. The two signatures intersect via two special predicates that form the core of an ALP. One, written $do(a)$, represents the execution of an action a by the agent, and the other, written $?(\phi)$, denotes the condition ϕ on the state of the environment in which the agent lives.

Example: An ALP for Peg Solitaire

Recall the Peg Solitaire domain from the previous chapter. In Section 4.1, we have seen that a general strategy for solving instances of this problem consists in successively identifying and solving specific, small patterns. The procedural action program written for this purpose required the use of additional fluents for the purpose of defining these patterns and other basic properties, like set membership. ALPs allow for a clear separation of the fluents and actions of a domain on the one hand, and the predicates needed to describe a strategy on the other hand. In the following, we give a declarative action program for Peg Solitaire using the fluent $Peg(x)$ to denote that cell x houses a peg, and action $Jump(x, y, z)$ to denote the action of jumping with the peg in x over the peg in y into cell z. To begin with, this ALP clause defines a predicate that is true if the board has been solved, which shall be the case when a single peg remains in the center square 45 on the board:

```
BoardSolved :- ?(peg(45) and forall(X, X=45 or not peg(X))).
```

The keywords `not`, `and`, `or`, and `forall` stand for the corresponding logical connectives.

As before, the solution strategy is based on the recognition of specific patterns. Recall that a pattern is a set of pegs in a specific arrangement along with a single "catalyst," as depicted in Figure 4.3(a). For the ALP, an instance of a pattern is encoded as a list of cells, the first of which denotes the position of the catalyst peg. The way to solve a pattern can then be defined as

```
patternSolved(P) :- member(Y,P), do(jump(X,Y,Z)), patternSolved(P).
```

Put in words, a pattern is solved by repeatedly jumping over a selected peg from the pattern until this situation is reached:

```
patternSolved([Catalyst|P]) :- ?(peg(Catalyst)), empty(P).
```

Put in words, the catalyst position is occupied while all other cells of the pattern, given in the tail list P, no longer contain a peg:

```
empty([]).
empty([X|L]) :- ?(not peg(X)), empty(L).
```

The solution of a single pattern is embedded in the definition of the overall solution strategy for the Peg Solitaire agent:

```
strategy :- boardSolved.
strategy :- isPattern(P), patternSolved(P), strategy.
```

Intuitively, the query `strategy` can be inferred from these clauses just in case there is a sequence of pattern-solving moves which leads from a given initial board configuration to a solved board. The program is completed by the encoding of suitable patterns, e.g., the ones depicted in Figure 4.3(a). For example, given the encoding of the cells as exemplified in Figure 5.1, the

		74	75	76		
	63	64	65	66	67	
52	53	54	55	56	57	58
42	43	44	45	46	47	48
32	33	34	35	36	37	38
	23	24	25	26	27	
		14	15	16		

FIGURE 5.1: An example encoding for the cells of a Peg Solitaire board.

3-pattern can be defined by the ALP clauses

```
isPattern([C,X1,X2,X3]) :- ?(peg(C)),
                           X1 = C+1, X2 = X1+10, X3 = X1+20,
                           ?(peg(X1) and peg(X2) and peg(X3)).

isPattern([C,X1,X2,X3]) :- ?(peg(C)),
                           X1 = C-1, X2 = X1+10, X3 = X1+20,
                           ?(peg(X1) and peg(X2) and peg(X3)).
```

Similar clauses can be used to encode the same pattern but with different orientation, as well as the other patterns.

The example ALP highlights the characteristics of declarative action programs: the syntax is similar to that of a standard logic program augmented by two special predicates linking the program to an underlying action domain. For the sake of simplicity, we consider only ALPs without negative body or query literals.

The syntax of agent logic programs over an action domain axiomatization Σ is defined as follows:

- The signature of the ALP includes all terms of sort FLUENT and ACTION from Σ.
- If p is an n-ary relation symbol that does not occur in the signature of Σ and $t_1, \ldots, t_n$ are terms, then $p(t_1, \ldots, t_n)$ is a program atom.
- If a is an ACTION term in Σ, then $do(a)$ is a program atom.
- If ϕ is a state property composed of FLUENT terms from Σ, then $?(\phi)$ is a program atom.
- If H, $B_1, \ldots, B_n$ are program atoms ($n \geq 0$), then H :- $B_1, \ldots, B_n$ is an *ALP clause* (with head H and body $B_1, \ldots, B_n$). If $n = 0$, this is simply written as H.
- An *agent logic program* is a finite set of ALP clauses.
- An *ALP query* is a finite sequence of program atoms $Q_1, \ldots, Q_n$.

5.2   ALP SEMANTICS

5.2.1   Declarative Semantics

ALPs combine logic programs, which are static in nature, with reasoning about actions, which describe changes over time. To understand its semantics, an ALP is expanded into a set of axioms in which the temporal aspect is made explicit. Specifically, every predicate $p(\vec{x})$ defined in the logic program is extended by two arguments of sort TIME, thus becoming $p(\vec{x}, s, t)$. So doing transforms the static predicates into dynamic ones, stipulating their temporal truth between two (possibly identical) time points. The special predicates, $\text{do}(a)$ and $?(\phi)$, are likewise extended by arguments of sort TIME. Together with the underlying domain axiomatization, these expanded clauses provide the declarative semantics of the ALP itself.

Formally, an ALP is expanded by expanding each of its clauses $H \mathbin{:-} B_1, \ldots, B_n$ (where $n \geq 0$) according to the following procedure. Let $s_1, \ldots, s_{n+1}$ be a sequence of TIME variables:

- For $i = 1, \ldots, n$,
    - If B_i is of the form $p(t_1, \ldots, t_m)$, it is expanded to $p(t_1, \ldots, t_m, s_i, s_{i+1})$;
    - If B_i is of the form $\text{do}(a)$, it is expanded to $Poss(a, s_i, s_{i+1})$;
    - If B_i is of the form $?(\phi)$, it is expanded to $\phi[s_i]$, and $s_{i+1} = s_i$.
    As usual, $\phi[s]$ stands for formula ϕ with all occurrences of a fluent f replaced by the atom $Holds(f, s)$.
- The head $H = p(t_1, \ldots, t_m)$ is expanded to $p(t_1, \ldots, t_m, s_1, s_{n+1})$.
- The symbols ":-" and "," are replaced by the implication "⊂" and the conjunction "∧", respectively.

For example, the two clauses describing the strategy in the program for Peg Solitaire are expanded to the implications

$$Strategy(s_1, s_2) \subset BoardSolved(s_1, s_2)$$
$$Strategy(s_1, s_4) \subset IsPattern(p, s_1, s_2) \land PatternSolved(p, s_2, s_3) \land Strategy(s_3, s_4)$$

The meaning of the clause that defines a solved board is given by

$$BoardSolved(s, s) \subset Holds(Peg(45), s) \land (\forall x)(x = 45 \lor \neg Holds(Peg(x), s))$$

It is important to realize that, unlike in standard logic programs, the order in which atoms occur in the body of a clause is crucial and needs to be respected when adding the time arguments to the individual atoms. Thus an ALP clause like, say,

```
p :- ?(f), do(a).
```

differs considerably from the clause

```
p :- do(a), ?(f).
```

In the first case, fluent f must hold prior to the execution of action a, whereas in the second clause the test applies to the situation after the action.

Queries to ALPs are expanded from left to right exactly like the body of a clause. The only difference is that the first TIME argument of the first atom is the initial timepoint of the time structure in the underlying action domain axiomatization. Thus the query strategy, for example, translates to the question whether the formula $(\exists s)\, Strategy(S_0, s)$ is logically entailed by the expanded program along with the underlying domain theory.

Time Structures for ALPs

The expansion of an ALP does not assume a particular time structure, so that the underlying action domain may use either linear or branching time. This is another difference to the procedural programming language GOLOG, whose semantics is closely coupled with the branching Situation Calculus. This notwithstanding, the expansion of an agent logic program does rely on a few assumptions about the underlying action domain regarding the time.

First and foremost, the execution of actions in ALPs is strictly sequential. Each atomic do(a) is expanded into $Poss(a, s, t)$ with s the initial and t the resulting timepoint. The underlying assumption, therefore, is that time progresses when an action is performed and that actions never overlap. Formally, the axiomatization of time in the underlying domain must entail

$$Poss(a, s, t) \supset s < t$$

and

$$Poss(a, s, t) \wedge Poss(a', s', t') \supset (t < t' \supset t \leq s') \wedge (t = t' \supset a = a \wedge s = s')$$

This condition is always satisfied, for instance, in a domain axiomatization based on situations where all precondition axioms are of the form

$$Poss(a, s, t) \equiv t = Do(a, s) \wedge \ldots$$

and where the standard foundational axioms on the ordering of situations hold (cf. (3.13) in Section 3.5).

A second assumption about the executability of actions is made when expanding sequences of actions by identifying the timepoint after an action with the initial timepoint of the following action, as in this expansion of the clause body do(a), do(b):

$$Poss(A, s_1, s_2) \wedge Poss(B, s_2, s_3) \tag{5.1}$$

This assumes that possible actions can begin right after a previous action has finished. If a domain violates this assumption, then queries which are intuitively true do in fact not follow from the expanded program. As an example, consider the two precondition axioms

$$Poss(A, s, t) \equiv s = 0 \wedge t = 1$$
$$Poss(B, s, t) \equiv s = 2 \wedge t = 3 \tag{5.2}$$

along with the ALP clause p :- do(a), do(b). Intuitively, it should be possible to do action A followed by action B and hence to derive query p; however, because A ends at time 1 and B cannot start before time 2, the expanded clause body, (5.1), does not follow from the domain axiomatization (5.2). This cannot happen if the background action theory satisfies the assumption that possible actions can always follow immediately after each other; formally,

$$(\exists t)\, Poss(a, s, t) \,\wedge\, (\forall f)\,(Holds(f, s) \equiv Holds(f, s')) \,\supset\, (\exists t')\, Poss(a, s', t')$$

Put in words, if the environment is in the same state at times s and s', then every action that is possible at s is also possible at s'. Under this assumption it is justified to link action execution in the way the expansion of an ALP does. This condition, too, is satisfied in every standard situation-style action domain axiomatization, but it is also worth realizing that this restriction can be lifted by modifying the definition of expansion in such a way that the starting point of an action is not identified with the termination of the preceding action.

5.2.2 Operational Semantics

In addition to the logical semantics, an operational semantics can be given for ALPs in the form of a proof calculus for query answering. Computing with an ALP requires to combine the standard derivation technique for logic programs with a reasoner for the underlying action domain axiomatization. The latter is needed to compute the special program atoms that refer, respectively, to the execution of an action, do(a), and to properties of the environment of the agent, $?(\phi)$.

Consider an expanded query Q to an expanded ALP P, and let Σ be the axiomatization of the underlying action domain. As an adaptation from the usual derivation technique for logic programs, the proof calculus establishes that $\Sigma \cup P \models Q\theta$ for some substitution θ, by systematically resolving the atoms in the query against applicable program clauses. The two special atoms are evaluated against Σ. Hence, the proof calculus relies on an appropriate derivation mechanism for the action domain, which takes care of the inference steps $\Sigma \models Poss(a, s, t)$ and $\Sigma \models \phi[s]$. Beginning with the given query, each derivation step maps a query to a new one, and a *successful derivation* ends with the empty query. A derived state indicates a failed derivation if none of the derivation rules applies. The derivation rules are as follows.

- *Standard Atoms*:

$$\frac{Q_1 \wedge Q_2 \wedge \ldots \wedge Q_n}{(B_1 \wedge \ldots \wedge B_m \wedge Q_2 \wedge \ldots \wedge Q_n)\theta}$$

where Q_1 is a literal defined in the ALP and $H \subset B_1 \wedge \ldots \wedge B_m$ is the expansion of a clause in the ALP such that $Q_1\theta = H\theta$, for some substitution θ.

- *Actions*:

$$\frac{Poss(a, s, t) \wedge Q_2 \wedge \ldots \wedge Q_n}{(Q_2 \wedge \ldots \wedge Q_n)\theta}$$

where $\Sigma \models Poss(a, s, t)\theta$ with substitution θ on the variables in $Poss(a, s, t)$.

- *Tests*:

$$\frac{\phi[s] \wedge Q_2 \wedge \ldots \wedge Q_n}{(Q_2 \wedge \ldots \wedge Q_n)\theta}$$

where $\Sigma \models \phi[s]\theta$ with substitution θ on the variables in ϕ.

If a derivation is successful, then the substitutions used in each derivation step can be combined and then restricted to the variables in the original queries. The resulting substitution is the *computed answer*.

To illustrate the use of these derivation rules, consider the program

```
office(alice, 101).
office(bob,   102).


deliver :- ?(hasPackageFor(P)), office(P,R), do(go(R)).
```

along with the query `deliver`. The expansion of the program is

Office(*Alice*, 101, s, s)
Office(*Bob*, 102, s, s)
Deliver(s_1, s_3) $\subset$ *Holds*(*HasPackageFor*(p), s_1) $\wedge$ *Office*(p, r, s_1, s_2) $\wedge$ *Poss*(*Go*(r), s_2, s_3)

Suppose the background theory Σ includes the fact *Holds*(*HasPackageFor*(*Bob*), S_0) along with the precondition axiom

$$Poss(Go(r), s, t) \equiv t = Do(Go(r), s)$$

Figure 5.2 depicts a successful derivation for the expanded query, *Deliver*(S_0, s), based on this background theory. The substitutions used in the individual derivation steps are, $\theta_1 = \{s_1/S_0, s_3/s\}$, $\theta_2 = \{p/Bob\}$ (following from the background theory), $\theta_3 = \{r/102, s_2/S_0\}$, and

$$\frac{\begin{array}{c} Deliver(S_0, s) \\ \hline Holds(HasPackageFor(p), S_0) \land Office(p, r, S_0, s_2) \land Poss(Go(r), s_2, s) \\ \hline Office(Bob, r, S_0, s_2) \land Poss(Go(r), s_2, s) \\ \hline Poss(Go(102), S_0, s) \end{array}}{\Box}$$

FIGURE 5.2: A derivation for an (expanded) ALP query.

$\theta_4 = \{s / Do(Go(102), S_0)\}$ (which again is computed from the background theory). The last replacement for variable s of the original query is also the computed answer.

Under the condition that the reasoner for the underlying action theory is correct, the derivation mechanism for ALPs is semantically correct. That is to say, if there exists a successful derivation for a query Q with computed answer θ, then the expanded program and the background theory together entail $Q\theta$. The proof calculus is in general not complete, however, if the agent has incomplete knowledge according to the underlying domain axiomatization. To illustrate this, consider the simple program

```
p :- ?(f).
p :- ?(not f).
```

Its expansion is obviously equivalent to the formula $P(s, s) \equiv True$, which entails that $(\exists s) P(S_0, s)$. However, there is no successful derivation for this query if the agent does not know whether or not fluent f holds initially.

5.3   AN ALP INTERPRETER

This section gives a brief introduction into the use of logic programming itself to implement a reasoner that is suitable for being combined with an ALP. As in case of the interpreter for GOLOG in Section 3.5, this includes a specific instance of an action calculus to specify a background theory as part of the logic program.

The *Fluent Calculus* is a variant of the Situation Calculus which uses the same branching time structure but adds a sort STATE as an explicit representation for states. Intuitively, a state is identified with the fluents that hold in it. The state in situation s is denoted by the standard function $State(s)$. By definition, each fluent itself is a (singleton) state, and if z_1 and z_2 are states, then so is their composition denoted by $z_1 \circ z_2$. The empty state is represented by the special constant $\emptyset$. By convention, STATE variables in Fluent Calculus axiomatizations are always denoted by z, possibly with subscripts or superscripts. The behavior of the function "$\circ$" is governed by the following foundational axioms, which essentially define states as non-nested sets of fluents:

$(z_1 \circ z_2) \circ z_3 = z_1 \circ (z_2 \circ z_3)$ $\qquad\qquad$ $z_1 \circ z_2 = z_2 \circ z_1$

$\neg Holds(f, \emptyset)$ $\qquad\qquad\qquad\qquad\qquad$ $Holds(f_1, f) \supset f_1 = f$

$Holds(f, z_1 \circ z_2) \supset Holds(f, z_1) \lor Holds(f, z_2)$ $\quad$ $(\forall f)(Holds(f, z_1) \equiv Holds(f, z_2)) \supset z_1 = z_2$

$(\forall P)(\exists z)(\forall f)(Holds(f, z) \equiv P(f))$

Here, $Holds(f, z)$ is used as an abbreviation for the equational formula $(\exists z')\, z = f \circ z'$, which amounts to an axiomatic characterizations of set membership. The last axiom, where P is a second-order predicate variable of sort FLUENT, stipulates the existence of a state for any (possibly infinite) set of fluents. These axioms are accompanied by the axiomatization of branching time as in the Situation Calculus (axioms (3.13) in Chapter 3).

The explicit notion of a state allows a simple and natural specification of positive and negative effects of actions, by a purely axiomatic characterization of subtraction and addition of fluents from and to states:

$$
\begin{aligned}
z_2 = z_1 + f &\overset{\text{def}}{=} z_2 = z_1 \circ f \\
z_2 = z_1 - f &\overset{\text{def}}{=} (z_2 = z_1 \lor z_2 + f = z_1) \land \neg Holds(f, z_2)
\end{aligned}
\tag{5.3}
$$

These macros can be straightforwardly generalized to the subtraction and addition of finitely many fluents.

Using the explicit concept of a state, action domains are axiomatized in the Fluent Calculus as follows, where the expression $Holds(f, s)$ in state formulas is now used as a mere abbreviation for $Holds(f, State(s))$. Precondition axioms, one for every action $A(\vec{x})$, are of the form

$$
Poss(A(\vec{x}), s) \equiv \pi_A[s]
\tag{5.4}
$$

where $\pi_A[s]$ is a state formula in s with free variables among $s, \vec{x}$. As in the Situation Calculus, the understanding is that an action always ends in the successor situation $Do(A(\vec{x}), s)$.

The effects of actions are specified by *state update axioms*, one for every action $A(\vec{x})$, which are of the form

$$
\begin{aligned}
Poss(A(\vec{x}), s) \supset (\exists \vec{y}_1)\,(\Phi_1[s] \land State(Do(A(\vec{x}), s))) &= State(s) - \vartheta_1^- + \vartheta_1^+) \\
\lor \ldots \lor & \\
(\exists \vec{y}_n)\,(\Phi_n[s] \land State(Do(A(\vec{x}), s))) &= State(s) - \vartheta_n^- + \vartheta_n^+)
\end{aligned}
\tag{5.5}
$$

where each $\Phi_i[s]$ is a state formula in s with free variables among $s, \vec{x}, \vec{y}_i$; and ϑ_i^- (the negative effects) and ϑ_i^+ (the positive effects) stand for zero or more subtractions and additions, respectively, of fluent terms with variables among $\vec{x}, \vec{y}_i$. A crucial property of the equations between a state and its successor is that the updated state always contains the positive effects plus all fluents in the preceding state except for the negative effects. This follows from the foundational axioms of the Fluent Calculus and definition (5.3), which entail that

$$State(Do(a,s)) = State(s) - g_1 - \cdots - g_m + f_1 + \cdots + f_n$$

implies

$$Holds(f, State(Do(a,s))) \equiv [\bigvee_i f = f_i] \vee [Holds(f, State(s)) \wedge \bigwedge_j f \neq g_j]$$

With this fundamental property of the Fluent Calculus, state update axioms can be understood as special instances of general effect axioms. Specifically, a state update axiom (5.5) corresponds to the effect axiom

$$Poss(A(\vec{x}), s, t) \supset$$
$$(\exists \vec{y}_1)(\Phi_1[s] \wedge (\forall f)[\bigvee_i f = f_{1i} \vee Holds(f,s) \wedge \bigwedge_j f \neq g_{1j} \supset Holds(f,t)]$$
$$\wedge (\forall f)[\bigvee_j f = g_{1j} \vee \neg Holds(f,s) \wedge \bigwedge_i f \neq f_{1i} \supset \neg Holds(f,t)])$$
$$\vee \ldots \vee$$
$$(\exists \vec{y}_n)(\Phi_n[s] \wedge (\forall f)[\bigvee_i f = f_{ni} \vee Holds(f,s) \wedge \bigwedge_j f \neq g_{nj} \supset Holds(f,t)]$$
$$\wedge (\forall f)[\bigvee_i f = g_{nj} \vee \neg Holds(f,s) \wedge \bigwedge_i f \neq f_{ni} \supset \neg Holds(f,t)])$$

Here, the f_{ki} and g_{kj} are the fluent terms that occur in ϑ_k^+ and ϑ_k^-, respectively. The reader may verify that this formula is equivalent to the state update axiom provided the latter is coherent, that is, fluents never occur as both positive and negative effect in the same update equation.

As an example of how to use the Fluent Calculus to specify an action domain, recall the one-player game from the beginning of Chapter 4 (cf. Figure 4.1). As a game with full information, its initial state is given by a complete collection of fluents.

$$State(S_0) = Cell(a, 1) \circ Cell(b, 1) \circ \ldots \circ Cell(h, 1) \tag{5.6}$$

The conditions for action $Jump(x, y)$, that is, to jump with the coin on position x onto the coin in place y, are formalized by the axiom

$$Poss(Jump(x, y), s) \equiv Holds(Cell(x, 1), s) \wedge Holds(Cell(y, 1), s) \wedge$$
$$CoinsBetween(x, y, s) = 2$$

Again, auxiliary function $CoinsBetween(x, y, s)$ counts the number of coins that lie between position x and y in $State(s)$. The following state update axiom formalizes the effect of this action:

$$Poss(Jump(x, y), s) \supset State(Do(Jump(x, y), s)) = State(s) - Cell(x, 1) - Cell(y, 1)$$
$$+ Cell(x, 0) + Cell(y, 2)$$

The basic version of the Fluent Calculus-based interpreter for agent logic programs is suitable for settings where agents have complete information about the state of their environment. Under this assumption, the state of the environment at any time during the computation is a finite collection of ground fluents, as in (5.6). These states are encoded by lists of fluents in the ALP interpreter, and a state is updated by standard operations on lists. To this end, the following basic and generic logic program uses the standard membership predicate, written member(X,Y), for atomic *Holds* statements, on the basis of which compound state formulas can be evaluated, too. The following program also uses a ternary version member(X,Y,Z) which means that X occurs in Y and Z is Y without X:

```
holds(F,Z) :- member(F,Z).


minus(Z,[],Z).
minus(Z,[F|Fs],Z2) :- member(F,Z,Z1), minus(Z1,Fs,Z2).
minus(Z,[F|Fs],Z2) :- not member(F,Z), minus(Z,Fs,Z2).


plus(Z,[],Z).
plus(Z,[F|Fs],Z2) :- not member(F,Z), plus([F|Z],Fs,Z2).
plus(Z,[F|Fs],Z2) :- member(F,Z), plus(Z,Fs,Z2).


update(Z1,ThetaP,ThetaN,Z2) :- minus(Z1,ThetaN,Z), plus(Z,ThetaP,Z2).
```

With the help of these basic definitions, the axioms of a concrete agent domain can be implemented straightforwardly. As an example, consider the encoding of the initial state along with the precondition and state update axioms in the single-player game from above:

```
init(Z0) :- Z0=[cell(a,1),cell(b,1),cell(c,1),cell(d,1),
               cell(e,1),cell(f,1),cell(g,1),cell(h,1)].


poss(jump(X,Y),Z) :- holds(cell(X,1),Z), holds(cell(Y,1),Z),
                 between(X,Y,Z,2).


do(Z1,S,jump(X,Y),Z2,do(jump(X,Y),S)) :-
    update(Z1,[cell(X,0),cell(Y,2)],[cell(X,1),cell(Y,1)],Z2).
```

The last axiom illustrates how the notion of a state allows us to maintain an explicit state in addition to the situation term. This method is known as *progression*, where the world model, i.e., the state, is updated whenever an action is performed. The advantage of so doing is that the updated state can be used directly for the evaluation of state properties. This is in contrast to the regression method used in the interpreter for GOLOG in Section 3.5, where state properties are rolled back to the initial situation prior to evaluation.

As a logic program, the background theory of an agent domain can be combined with the ALP itself. If the Fluent Calculus is used with its duality of states and situations, then the expanded ALP can use this pair, too, in the place of a single timepoint. This is illustrated in the following encoding of an (expanded) ALP that defines a complete search for the Coin game:

```
strategy(S) :- init(Z0), plan(Z0,s0,Z,S).


plan(Z,S,Z,S) :- goal(Z).
plan(Z1,S1,Z3,S3) :- poss(A,Z1), do(Z1,S1,A,Z2,S2),
                     plan(Z2,S2,A,Z3,S3).


goal(Z) :- holds(not exists(X, cell(X,1)),Z).
```

According to this definition, a goal state is reached if no single coin is left, which is equivalent to the requirement that all stacks have two coins. Among the answers to the query `strategy(S)` is

```
S = do(jump(e,h),do(jump(a,c),do(jump(f,b),do(jump(d,g),s0)))))
```

5.4   AGENT LOGIC PROGRAMS WITH SENSING

The encoding of states as ground lists of fluents in the basic ALP interpreter is suitable only for domains in which the agent has complete information about its environment. A more sophisticated state representation is required in the general case of incomplete knowledge and if agents not only act but also sense.

Sensing actions can be modeled in declarative action programs in much the same way as described in Section 3.7 for procedural action programs. For the sake of simplicity, we again consider binary sensing actions only, that is, which allow the agent to learn the truth-value of a specific fluent. Sensing actions and their outcomes can be integrated into an ALP by using, as before, the concept of a sensing history. Recall that the expressions $SF(a, s)$ and $Sensed[h, s]$ denote domain-specific formulas that encode, respectively, the effect of sensing action a and the entire sensing information given in history h beginning with situation s.

To compute ALPs which include sensing actions, the histories play a role when inferring the special atoms. To this end, the two derivation rules for these cases are generalized as follows:

- *Actions*:

$$\frac{Poss(a, s, t) \wedge Q_2 \wedge \ldots \wedge Q_n}{(Q_2 \wedge \ldots \wedge Q_n)\theta}$$

where $\Sigma \cup \{Sensed[h, s]\} \models Poss(a, s, t)\theta$ with substitution θ on the variables in $Poss(a, s, t)$.

- *Tests*:

$$\frac{\phi[s] \wedge Q_2 \wedge \ldots \wedge Q_n}{(Q_2 \wedge \ldots \wedge Q_n)\theta}$$

where $\Sigma \cup \{Sensed[h, s]\} \models \phi[s]\theta$ with substitution θ on the variables in ϕ.

The encoding of states in the ALP interpreter can be extended to incomplete knowledge with the help of *constraint logic programming*. To this end, incomplete states are encoded by open-ended lists of fluents, now possibly containing variables,

```
Z = [F1,...,Fk | Z1 ]
```

where the tail Z1 is a variable indicating that more fluents may hold but which are, at present, unknown. If the arguments of all fluents are encoded by numbers, then a standard arithmetic solver can be used for constraints on partially known arguments. Negative and disjunctive state knowledge is then expressed by additional *state constraints*. A basic set of these constraints is depicted in Figure 5.3. While a variety of state formulas can be encoded with the help of just these three, they do not cover full first-order logic, and specific applications may require additional or other constraints. An auxiliary constraint, written duplicate_free(Z), is needed to stipulate that a list of fluents contains no multiple occurrences of the same fluent.

Constraint	Meaning
not_holds(F,Z)	$\neg Holds(f, z)$
not_holds_all(F,Z)	$(\forall \vec{x}) \neg Holds(f, z)$, $\vec{x}$ variables in f
or_holds([F1,...,Fn],Z)	$\bigvee_{i=1}^{n} Holds(f_i, z)$

FIGURE 5.3: Constraints to express atomic negative state knowledge, universally quantified negations, and disjunctions.

The ALP interpreter described in the previous section can be extended by a method to handle the state constraints using a general technique of specifying, in a declarative way, rules for processing constraints.

A *constraint handling rule* is of the form

$$H_1, \ldots, H_m \Leftrightarrow G_1, \ldots, G_k \mid B_1, \ldots, B_n \tag{5.7}$$

where

- the *head* $H_1, \ldots, H_m$ is a sequence of constraints $(m \geq 1)$;
- the *guard* $G_1, \ldots, G_k$ and the *body* $B_1, \ldots, B_n$ are queries $(k, n \geq 0)$.

An empty guard is omitted; the empty body is denoted by *True*. The declarative interpretation of such a rule is given by the formula

$$(\forall \vec{x})(G_1 \wedge \ldots \wedge G_k \supset [H_1 \wedge \ldots \wedge H_m \equiv (\exists \vec{y})(B_1 \wedge \ldots \wedge B_n)]) \tag{5.8}$$

where $\vec{x}$ are the variables in both guard and head and $\vec{y}$ are the variables which additionally occur in the body.

The procedural interpretation of a constraint handling rule is given by a transition in a constraint store: if the head can be matched against elements of the constraint store and the guard can be derived, then the constraints which match the head are replaced by the body.

Figure 5.4 depicts suitable constraint handling rules for some of the basic state constraints. These can be extended by rules for the disjunctive constraint as well as for other constraints that constitute the elements for encoding state knowledge in a tailor-made interpreter for ALPs. Based on their declarative interpretation, the correctness of a set of constraint handling rules can be proved against the foundational axioms of the Fluent Calculus.

```
not_holds(_,[])          <=> true.
not_holds(F,[F1|Z])      <=> neq(F,F1), not_holds(F,Z).

duplicate_free([])       <=> true.
duplicate_free([F|Z])    <=> not_holds(F,Z), duplicate_free(Z).
```

FIGURE 5.4: Constraint handling rules for the negation constraint and the constraint on multiple occurrences of fluents. Auxiliary constraint $neq(f_1, f_2)$ means the inequality of fluents f_1 and f_2.

The incorporation of constraints and constraints handling rules into the interpreter for ALPs allows us to write control programs for agents with incomplete knowledge of their environment. In each computation step, the current world model is given by an incomplete list of fluents plus a set of constraints. Tests are then evaluated against this incomplete model, and the state encoding is updated to a new list and a modified set of constraints whenever an action is performed. If the ALP includes the use of sensing actions, then whenever the agent learns the truth-value of a state property, this information is incorporated into the current state description.

Example: Exploration Agent

Consider the model of an adventure game depicted in Figure 5.5. An agent is placed somewhere in an environment of unknown size. Its task is to find as many gold items as possible and to take them to the depot. The agent must, however, avoid falling into any of the pits. The agent is equipped with a sensor which tells it whether it is in a location next to a pit but without revealing the direction in which the pit actually is. A second sensor is activated whenever the agent is in a cell containing gold. Besides sensing, the basic actions of the agent shall be to move to the adjacent cell in any direction and to pick and drop gold. A suitable control program should have the agent systematically explore the environment to collect gold while avoiding to enter a square with a pit. This problem illustrates two challenges raised by incomplete state knowledge: agents have to act cautiously, and they need to interpret and logically combine sensor information acquired over time.

The formal description of this problem must include the effects of the two sensing actions with the help of the predicate $SF(a, s)$. Let the fluent $Gold(x, y)$ describe the presence of a

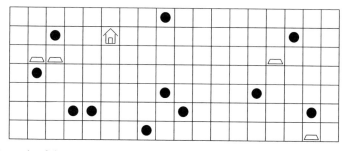

FIGURE 5.5: The task of the agent in this world is to explore the initially unknown environment with the goal to find and collect as many of the four gold nuggets as it can safely get without risking to fall into any of the circular pits. While the agent can sense the presence of a pit if standing next to one, it is unable to tell in which direction the pit lies. This actually makes it impossible to safely collect the leftmost gold nugget in the particular scenario depicted here.

gold nugget in cell (x, y), then the action $SenseGold(x, y)$ is suitably described by the axiom

$$SF(SenseGold(x, y), s) \equiv Holds(Gold(x, y), s)$$

Slightly more involved is the definition of the effect of sensing a pit, which means that at least one of the four squares around the agent houses a pit:

$$SF(SensePit(x, y), s) \equiv Holds(Pit(x + 1, y), s) \vee Holds(Pit(x, y + 1), s) \vee$$
$$Holds(Pit(x - 1, y), s) \vee Holds(Pit(x, y - 1), s)$$

In the ALP interpreter, these two axioms can be encoded with the help of the state constraints from Figure 5.3 as follows, where the acquired sensor information, true or false, is evaluated against a given, incomplete state Z:

```
sf(senseGold(X,Y),true,Z)   :- holds(gold(X,Y),Z).
sf(senseGold(X,Y),false,Z)  :- not_holds(gold(X,Y),Z).

sf(sensePit(X,Y),true,Z) :-
    X_east=X+1, X_west=X-1, Y_north=Y+1, Y_south=Y-1,
    or_holds([pit(X_east,Y),pit(X,Y_north),
              pit(X_west,Y),pit(X,Y_south)],Z).

sf(sensePit(X,Y),false,Z) :-
    X_east=X+1, X_west=X-1, Y_north=Y+1, Y_south=Y-1,
    not_holds(pit(X_east,Y),Z), not_holds(pit(X,Y_north),Z),
    not_holds(pit(X_west,Y),Z), not_holds(pit(X,Y_south),Z).
```

Whenever the agent makes an observation and incorporates it into the state description, the constraint solving mechanism combines the new information with what the agent already knows. For example, suppose the current incomplete state is given by the incomplete list Z=[at(3,5)|Z1] (that is, the only fluent known to be true is the agent's position) along with the two constraints not_holds(pit(3,4),Z1) and not_holds(pit(4,5),Z1). If in this situation the agent senses both gold and the presence of a nearby pit, then solving the two atoms

```
sf(senseGold(3,5),true,Z),
sf(sensePit(3,5), true,Z)
```

results in the substitution {Z1/[gold(3,5)|Z2}} and gives rise to the extended state description

```
Z=[at(3,5),gold(3,5) | Z2],
not_holds(pit(3,4),Z2),
not_holds(pit(4,5),Z2),
or_holds([pit(3,6),pit(2,5)],Z2).
```

The last disjunction reflects the fact that the agent is uncertain about whether a pit is to the left or above the current location (or both). This can only be further resolved if later and in another square, say (1,5), the agent's sensor would not be activated, implying that actually it must be cell (3,6) that houses a pit.

The specification of the sensing actions of the agent can be extended by suitable precondition and state update axioms for the other actions. The resulting logic program can then be combined with an arbitrary ALP acting as the actual control program for the agent.

5.5   EXERCISES

5.1. Axiomatize an elevator control scenario under the assumption that the destinations of all passengers are known initially! Write an ALP which tries to minimize the number of stops for people inside the elevator!

5.2. (a) Prove that the foundational axioms of the Fluent Calculus entail each of the following:

- $z \circ z = z$
- $\bigwedge_{i=1}^{n} Holds(f_i, z) \supset z + f_1 + \cdots + f_n = z$
- $\bigwedge_{i=1}^{n} \neg Holds(f_i, z) \supset z - f_1 - \cdots - f_n = z$

   (b) Show that the foundational axioms of the Fluent Calculus are mutually independent, by finding a model for the negation of each axiom with all other axioms satisfied!

5.3. Axiomatize the Knight's Tour Problem for boards of various sizes; see Figure 5.6! Find a good solution strategy for this problem (an example could be a heuristics by which the "mobility" of the knight is maximized in every situation) and write an ALP that implements a solution strategy for this problem! Run this ALP using the logic program from Section 5.3!

5.4. Modify the elevator control axiomatization from Exercise 5.1 as follows. The destinations are not known initially; instead, people press a button when they enter the

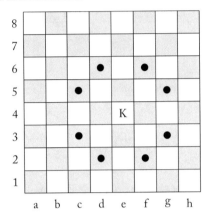

FIGURE 5.6: The Knight's Tour Problem: a knight is placed somewhere on a chess board. The task is to find a sequence of knight moves by which each square is visited exactly once. For illustration, the eight possible moves of a knight on e4 are depicted. A knight at the border or in a corner has similar but fewer moves. For a simpler variant of the problem, assume that the left and right hand side of the board as well as the top and bottom are connected, so that the knight has eight moves from all squares.

elevator. Furthermore, the capacity of the elevator is limited. Write an ALP and implement it using the constraint logic program from Section 5.4!

5.5. Write and implement an ALP for the gold mining agent to find and collect as many gold items as possible in scenarios such as the one depicted in Figure 5.5! Assume that initially the agent only knows that it starts at the depot and that the four cells that border this base location are safe.

CHAPTER 6

Reactive Action Programs

A procedural or a declarative action program, possibly coupled with planning, provides a complex and long-term strategy for an autonomous agent. These approaches are suited for agents that live in a structured, reasonably well-behaved world, over which they have sufficient control. In environments which are highly dynamic and much less controllable, however, elaborate strategies and plans are often inappropriate. Frequent and unforeseeable changes require flexible control programs, which allow agents to constantly assess their current strategy and to quickly adapt when necessary. Software agents in an open world or autonomous robots in unfamiliar environments are examples of systems for which a sophisticated strategy is unsuited because it is likely that they are unable to complete any long-term behavior without the need to adapt to unexpected circumstances. Under these conditions, reactive action programs are often a better choice. As a form of behavior-based control, reactive programs require the agents to constantly choose short-term goals which can be easily achieved but also quickly abandoned in case of unexpected changes in the environment.

6.1 BDI-AGENTS

Instead of following a single, elaborate strategy, a behavior-based agent has at its disposal a variety of comparably simple behaviors (usually called *procedures* in this context) which they select according to their current needs. A single procedure is meant to achieve a short-term goal, and agents can quickly abandon a chosen behavior and adopt a new one if necessary. Programming a reactive agent thus requires to provide these short-term procedures along with suitable criteria under which they can and should be adopted. The standard approach to this form of agent programming is the *BDI-model*, where the capitals stand for Belief, Desire, and Intention.

In the BDI-model, the state of an agent at any time is characterized by three components.

- *Beliefs* constitute the internal world model: They describe what the agent currently believes about the environment and its own tasks.

- *Desires* are derived from the beliefs and describe what the agent currently tries to achieve.
- *Intentions* are the behaviors (procedures) which the agent has adopted and is currently following in order to meet its desires.

A *Procedural Reasoning System* (PRS) combines the state components of a BDI-based agent with a library of programmed procedures and a reasoner to control the behavior of the agent. Figure 6.1 depicts the basic architecture of a PRS containing these modules along with a connection to the outside world through the sensors and effectors of an agent. Being a general and abstract framework for reactive action programs, the BDI-model can be instantiated with different languages for the specification of beliefs and desires and for the programming of procedures.

Beliefs

The database containing the current beliefs of an agent during the execution of a reactive action program is the PRS-equivalent of the internal world model maintained in procedural or declarative programs. Hence, beliefs are held about the fluents which describe the relevant properties of the environment of the agent. As usual, the beliefs are affected by both the agent's actions and the acquired sensing information. Besides knowledge of fluents, a belief base may include static knowledge of the environment, which corresponds to domain constraints in general action calculi. A specialty of the state representation in reactive action programs is that it may also include properties which characterize the current behavioral stance of the agent itself. Like all dynamic state components, these "mental" fluents may be affected by both actions and sensing information, in which case they take the agent into a different state of mind.

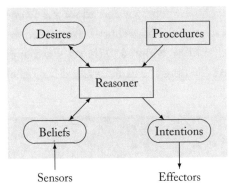

FIGURE 6.1: A general PRS-architecture for BDI-based agents.

Desires

Desires are formalized as properties of the environment that the agent currently wants to achieve. The simple form of a desire is a goal formula, indicating that the agent aims at a state in which the goal is satisfied. These correspond to goals in planning problems, but are typically short-term and an agent can have several of these desires at the same time. More expressive settings support temporally extended desires, such as the maintenance of a state property over a certain time interval, similar to the temporal logic formulas used in Chapter 4 for control formulas and preferences.

Intentions

Intentions are collections of instantiated procedures which the agent has selected in order to achieve its desires. Intentions can be active or passive, and only in the former case can the agent select its next action according to this intention. Typically, intentions are partially ordered, in which case only the intentions on top are active and all others remain passive until the preceding intentions have been successfully completed.

Procedures

The procedures constitute the behavioral knowledge of the agent about how to achieve goals or how to react under given circumstances. In the simple case, a procedure is a sequence of actions, while more expressive languages allow conditionals and loops as in a procedural action program or in an agent logic program. Procedures correspond to, typically short, GOLOG procedure definitions or to clausal definitions in an ALP. Each procedure is associated with a condition that defines the desires for which the behavior is suitable. A procedure may affect the behavioral stance of the agent if it includes actions which change the beliefs that encode this mental state.

Reasoner

The reasoner controls the execution of a reactive action program. Executing usually happens in form of a cycle, where sensing is followed by selection and then acting. Starting with an initial belief base, the reasoner derives the desires that follow from the beliefs and then chooses appropriate intentions for these desires. This is followed by selecting an active intention and then executing the next step of this behavior. This results in an updated set of beliefs according to the effect of the chosen action and possibly acquired sensing information, and then the cycle continues. The program ends when a belief state is reached which implies no more desires.

6.2   AGENTSPEAK

AgentSpeak is a simple programming language for reactive agents based on the BDI-model. It provides a syntax for encoding the belief base and for defining a set of procedures for an agent. The language admits a formal operational semantics which follows the general PRS-architecture. Borrowing many of the principles of logic programming, AgentSpeak can be easily implemented in a declarative fashion.

6.2.1   Syntax

A world model in AgentSpeak is composed of *belief atoms* representing the relevant properties of the environment. This includes both static and dynamic beliefs. The latter correspond to fluents and may be affected by the actions of the agent or otherwise change while the agent program is executed. Besides predicates for beliefs, a domain signature contains predicates representing the possible actions of the agent.

Definition 6.2.1. *An* AgentSpeak domain signature *includes a finite set of belief predicates and a finite set of action predicates. A* belief literal *is an atom with a belief predicate or its negation.*   □

The beliefs of an agent at any time are given by a set of variable-free atoms. For the purpose of inference, these atoms are treated as (dynamically changing) facts in a logic program.

Example: Reactive Robot

Consider an environment similar to the one depicted in Figure 5.5 but with less regular a structure and where instead of static pits there are moving obstacles which from time to time prevent the agent from entering a specific location. The environment shall be described with the help of the following belief predicates:

Symbol	Type
Adjacent	CELL × CELL
At	{*Agent, Gold, Obstacle, Depot*} × CELL
Carries	{*Agent*} × {*Gold*}

Thus the (static) layout is encoded by an adjacency relation for the locations, and the fluent *At* is used for the (static and dynamic) positions of the various objects and the agent itself. An initial database of beliefs may then be given by

$$Adjacent(A, B)$$
$$Adjacent(B, C)$$
$$At(Agent, A)$$
$$At(Gold, A)$$
$$At(Depot, C)$$

Suppose further that the agent has at its disposal the following three action predicates:

Symbol	Type	Meaning
Pick	{*Gold*}	pick up a gold item
Drop	{*Gold*}	drop a gold item
Move	CELL × CELL	move to an adjacent cell

In order to keep the definition of individual behaviors conceptually simple, behaviors in AgentSpeak are mere sequences of elementary steps. Such a step can be an action of the agent or the addition of a new goal. There are two kinds of goals: achieving a specific property, written $!f$, where f is a belief atom; or establishing that a property holds, written $?f$. Sequences of actions and goals are treated like queries in a logic program. An example of such a sequence is

$$Pick(Gold), \ !At(Agent, y), \ Drop(Gold) \tag{6.1}$$

This behavior, when adopted for some given y, causes the agent to pick up gold, then to set the goal to be at location y, and finally to drop the gold after this goal has been achieved. Another example of a behavior is

$$?At(Gold, x), \ !At(Agent, x)$$

This sequence requires the agent to find an instance for location x of which it believes that gold can be found there, and then to set the goal to be at this x.

Each procedure definition needs to be accompanied by a condition under which the specified behavior can be adopted by the agent. This condition consists of two parts: a triggering event, which must have arisen and for which the sequence is appropriate, and a context defining preconditions for the applicability of the procedure. Triggering events can be the addition or removal of either a belief or a goal.

Definition 6.2.2. *If f is a belief atom, then $!f$ and $?f$ are goals. A triggering event is any of $+f$, $-f$, $+g$, $-g$, where f is a belief atom and g a goal.*

A procedure is an expression

$$e \ : \ b_1, \ldots, b_m \ \leftarrow \ p_1, \ldots, p_n$$

where e is a triggering event, $b_1, \ldots, b_m$ (the context) are belief literals ($m \geq 0$), and $p_1, \ldots, p_n$ (the body) is a sequence of action atoms or goals ($n \geq 0$). An empty context or body is denoted by True. $\square$

An example of a procedure for the gold mining agent is the following, which uses the three-step sequence (6.1) from above:

$$+At(Gold, x) : At(Agent, x), At(Depot, y) \ \leftarrow \ Pick(Gold), \ !At(Agent, y), \ Drop(Gold)$$

This procedure is triggered if the agent has learned that gold is in cell x, and the behavior can be adopted if the agent happens to be at x and the location y of the depot is known. The body of this procedure has the agent pick up the gold, then to set the goal to be at the depot, and finally to drop the gold. This behavior may be accompanied by two procedures that tell the agent what to do to safely reach a specific location:

$$+ \,!At(Agent, x) \,:\, At(Agent, x) \;\leftarrow\; True$$

$$+ \,!At(Agent, x) \,:\, At(Agent, y),\; x \neq y,\; Adjacent(y, z),\; \neg At(Obstacle, z) \quad (6.2)$$
$$\leftarrow\; Move(y, z),\; !At(Agent, x)$$

The first procedure says that the goal to be at some location is achieved if the agent is already there, and the second procedure says that the same goal can be achieved by a behavior that first takes a single step to an adjacent, currently unobstructed cell and then to pursue again the goal to reach x.

6.2.2   Operational Semantics

The generic PRS-architecture determines the underlying execution principle for agent programs written in AgentSpeak. With the help of the BDI-model, the state of the agent at any time is characterized by the following three sets:

- $\mathcal{B}$, a set of variable-free belief atoms.
- $\mathcal{I}$, a set of intentions. A single intention is an ordered set

$$[P_1; \ldots; P_k]$$

  where each P_i is a procedure body, possibly with some variables instantiated ($k \geq 0$). The first procedure, P_1, is always the one with the highest priority and thus the only one in the intention that is currently active.

- $\mathcal{D}$, a set of desires, each of which is of the form $\langle e; i \rangle$ where e is a triggering event and i an intention.

In general, the second component of a desire, i, encodes the remaining steps of the procedure which has generated event e. For example, executing the second step of sequence (6.1) may yield the desire

$$\langle +!At(Agent, C); [Drop(Gold)] \rangle$$

This indicates that the agent desires $At(Agent, C)$ in order to be able to continue with the procedure body $[Drop(Gold)]$. A special form are *external desires* given to the agent by its user.

These desires are formalized as

$$\langle e, [\,] \rangle$$

where e is a triggering event representing an external goal.

The three standard components of a BDI-agent are accompanied by these three selection functions:

- $\mathcal{S}_D$ selects an element from the current desires;
- $\mathcal{S}_I$ selects an element from the current intentions;
- $\mathcal{S}_P$ selects an applicable procedure for a triggering event.

Function $\mathcal{S}_P$ selects applicable procedure instances in accordance with the following definition:

Definition 6.2.3. *Consider a set $\mathcal{B}$ of variable-free belief atoms. Let e be a triggering event and P a procedure*

$$d \;:\; b_1, \ldots, b_m \;\leftarrow\; p_1, \ldots, p_n$$

Then P is relevant for e if $d\theta = e\theta$ for some substitution θ. If, furthermore, a substitution η exists such that $(b_1 \wedge \ldots \wedge b_m)\theta\eta$ is a logical consequence of $\mathcal{B}$, then the procedure instance $P\theta\eta$ is applicable to e (w.r.t. $\mathcal{B}$). □

Derivability follows the usual definition of logic programs, including the principle of negation-by-failure. For example, assuming the initial database from above which includes *At(Agent, A)* and *Adjacent(A, B)* but not *At(Obstacle, B)*, consider the triggering event $+!$ *At(Agent, C)*. While both procedures in (6.2) are relevant in this case, only the second one is applicable, determining the substitutions $\theta = \{x/C\}$ and $\eta = \{y/A, z/B\}$. The resulting behavior is then given by the instance

$$Move(A, B), \;!At(Agent, C)$$

Derivation Rules

The operational semantics of AgentSpeak is given by a set of derivation rules on agent configurations. At any time during the execution of an AgentSpeak program, the agent is characterized by a 4-tuple $(\mathcal{B}, \mathcal{D}, \mathcal{I}, \sigma)$ consisting of sets of beliefs, desires, and intentions, along with a parameter $\sigma \in \{\texttt{Sense}, \texttt{Select}, \texttt{Act}\}$ denoting the current stage in the sense-select-act cycle. The procedure definitions are not part of a configuration as they are assumed to be given and constant. The initial state is $(\mathcal{B}, \{\}, \{\}, \texttt{Sense})$, where $\mathcal{B}$ is an arbitrary initial set of belief atoms.

The only derivation rule for the stage in which the agent senses is

$$\frac{(\mathcal{B}, \mathcal{D}, \mathcal{I}, \texttt{Sense})}{(\mathcal{B}', \mathcal{D}', \mathcal{I}, \texttt{Select})}$$

where $\mathcal{B}'$ is obtained by updating $\mathcal{B}$ according to the sensing result and where $\mathcal{D}'$ is $\mathcal{D}$ augmented by all external desires sensed by the agent.

For the selection of a desire, four cases are distinguished. If there are no desires, the agent proceeds by acting according to one of its intentions. If the selected desire has no relevant procedure, it is dropped as the agent will never be able even to intend it. If there is a relevant procedure for the selected desire but none that is currently applicable, then the agent keeps the desire but continues with selecting an action for an existing intention. Finally, in case there are applicable procedures for a selected desire, a selected procedure instance is added to the intentions and the agent continues with choosing an action.

1. For $\mathcal{D} = \{\}$,

$$\frac{(\mathcal{B}, \{\}, \mathcal{I}, \texttt{Select})}{(\mathcal{B}, \{\}, \mathcal{I}, \texttt{Act})}.$$

2. If $\mathcal{S}_D(\mathcal{D}) = \langle e; i \rangle$ such that there is no relevant procedure for e, then

$$\frac{(\mathcal{B}, \mathcal{D}, \mathcal{I}, \texttt{Select})}{(\mathcal{B}, \mathcal{D} \setminus \{\langle e; i \rangle\}, \mathcal{I}, \texttt{Select})}.$$

3. If $\mathcal{S}_D(\mathcal{D}) = \langle e; i \rangle$ such that there is a relevant procedure for e but none that is applicable w.r.t. $\mathcal{B}$, then

$$\frac{(\mathcal{B}, \mathcal{D}, \mathcal{I}, \texttt{Select})}{(\mathcal{B}, \mathcal{D}, \mathcal{I}, \texttt{Act})}.$$

4. If $\mathcal{S}_D(\mathcal{D}) = \langle e; i \rangle$ and $\mathcal{S}_P(e) = P\theta\eta$, then
   (a) for external desires, where $i = [\,]$,

$$\frac{(\mathcal{B}, \mathcal{D}, \mathcal{I}, \texttt{Select})}{(\mathcal{B}, \mathcal{D} \setminus \{\langle e; [\,] \rangle\}, \mathcal{I} \cup \{[P\theta\eta]\}, \texttt{Act})};$$

   (b) for internal desires, where $i = [P_1; \ldots; P_k]$

$$\frac{(\mathcal{B}, \mathcal{D}, \mathcal{I}, \texttt{Select})}{(\mathcal{B}, \mathcal{D} \setminus \{\langle e; i \rangle\}, \mathcal{I} \cup \{[P\theta\eta; P_1; \ldots; P_k]\}, \texttt{Act})}.$$

For the last step in the sense-select-act cycle, if there are no intentions, then nothing changes and the agent simply starts the cycle again. Otherwise the next step in a selected intention is executed. If this is an action, then the actual execution of the action is reflected in an updated belief base according to the effects of that action. If it is an achievement goal $!f$, then a new desire is obtained with this goal as triggering event. If the goal is to establish a property, $?f$, then either this can be established from the current belief base, or a new desire is generated with triggering event $+?f$.

In case the executed step happens to be the last one in a procedure body, the next procedure in the intention stack becomes active, and if there is no further procedure, then the intention has been fully achieved. In any case, the sense-select-act cycle starts again after the execution of one step of an intention.

1. For $\mathcal{I} = \{\}$,

$$\frac{(\mathcal{B}, \mathcal{D}, \{\}, \mathsf{Act})}{(\mathcal{B}, \mathcal{D}, \{\}, \mathsf{Sense})}.$$

2. If $\mathcal{S}_I(\mathcal{I}) = [a, P_1; \ldots; P_k]$ with a an action, then

$$\frac{(\mathcal{B}, \mathcal{D}, \mathcal{I}, \mathsf{Act})}{(\mathcal{B}', \mathcal{D}, \mathcal{I} \setminus \{[a, P_1; \ldots; P_k]\} \cup \{[P_1; \ldots; P_k]\}, \mathsf{Sense})}$$

where $\mathcal{B}'$ is the result of updating $\mathcal{B}$ by the effects of action a.

3. If $\mathcal{S}_I(\mathcal{I}) = [!f, P_1; \ldots; P_k]$, then

$$\frac{(\mathcal{B}, \mathcal{D}, \mathcal{I}, \mathsf{Act})}{(\mathcal{B}, \mathcal{D} \cup \{\langle +!f; [P_1; \ldots; P_k] \rangle\}, \mathcal{I} \setminus \{[!f, P_1; \ldots; P_k]\}, \mathsf{Sense})}.$$

4. If $\mathcal{S}_I(\mathcal{I}) = [?f, P_1; \ldots; P_k]$ and θ is a substitution such that $\mathcal{B}$ entails $f\theta$, then

$$\frac{(\mathcal{B}, \mathcal{D}, \mathcal{I}, \mathsf{Act})}{(\mathcal{B}, \mathcal{D}, \mathcal{I} \setminus \{[?f, P_1; \ldots; P_k]\} \cup \{[P_1; \ldots; P_k]\theta\}, \mathsf{Sense})}.$$

5. If $\mathcal{S}_I(\mathcal{I}) = [?f, P_1; \ldots; P_k]$ while there is no θ such that $\mathcal{B}$ entails $f\theta$, then

$$\frac{(\mathcal{B}, \mathcal{D}, \mathcal{I}, \mathsf{Act})}{(\mathcal{B}, \mathcal{D} \cup \{\langle +?f; [P_1; \ldots; P_k] \rangle\}, \mathcal{I} \setminus \{[?f, P_1; \ldots; P_k]\}, \mathsf{Sense})}.$$

6. If $\mathcal{S}_I(\mathcal{I}) = [True; P_2; \ldots; P_k]$, then

$$\frac{(\mathcal{B}, \mathcal{D}, \mathcal{I}, \mathsf{Act})}{(\mathcal{B}, \mathcal{D}, \mathcal{I} \setminus \{[True; P_2; \ldots; P_k]\} \cup \{[P_2; \ldots; P_k]\}, \mathsf{Sense})}.$$

7. If $\mathcal{S}_I(\mathcal{I}) = [\,]$, then

$$\frac{(\mathcal{B}, \mathcal{D}, \mathcal{I}, \mathsf{Act})}{(\mathcal{B}, \mathcal{D}, \mathcal{I} \setminus \{[\,]\}, \mathsf{Sense})}.$$

For illustration, Figure 6.2 shows the initial steps of a derivation with an initial desire triggered by the fact that the agent in the gold mining scenario has just detected gold at location A.

$$\mathcal{B} = \{At(Agent, A), At(Gold, A), At(Depot, C)\}$$
$$\mathcal{D} = \{\langle +At(Gold, A); [\,]\rangle\}$$
$$\mathcal{I} = \{\}$$
$$\sigma = \texttt{Select}$$

$$\mathcal{B} = \{At(Agent, A), At(Gold, A), At(Depot, C)\}$$
$$\mathcal{D} = \{\}$$
$$\mathcal{I} = \{[Pick(Gold), !At(Agent, C), Drop(Gold)]\}$$
$$\sigma = \texttt{Act}$$

$$\mathcal{B} = \{At(Agent, A), Carries(Agent, Gold), At(Depot, C)\}$$
$$\mathcal{D} = \{\}$$
$$\mathcal{I} = \{[!At(Agent, C), Drop(Gold)]\}$$
$$\sigma = \texttt{Sense}$$

$$\mathcal{B} = \{At(Agent, A), Carries(Agent, Gold), At(Depot, C)\}$$
$$\mathcal{D} = \{\}$$
$$\mathcal{I} = \{[!At(Agent, C), Drop(Gold)]\}$$
$$\sigma = \texttt{Select}$$

$$\mathcal{B} = \{At(Agent, A), Carries(Agent, Gold), At(Depot, C)\}$$
$$\mathcal{D} = \{\}$$
$$\mathcal{I} = \{[!At(Agent, C), Drop(Gold)]\}$$
$$\sigma = \texttt{Act}$$

$$\mathcal{B} = \{At(Agent, A), Carries(Agent, Gold), At(Depot, C)\}$$
$$\mathcal{D} = \{\langle +!At(Agent, C); [Drop(Gold)]\rangle\}$$
$$\mathcal{I} = \{\}$$
$$\sigma = \texttt{Sense}$$

FIGURE 6.2: The initial steps in an AgentSpeak derivation. At the beginning, an appropriate instance of the applicable procedure for the external goal $+At(Gold, A)$ is selected. In the next step, the first action, picking up the gold, is executed and the beliefs are updated accordingly. The last step shown executes the goal $!At(Agent, C)$, giving rise to a new, internal desire. For the sake of simplicity, the static beliefs about the adjacency relation are not shown.

6.3   AN AGENTSPEAK INTERPRETER

Because AgentSpeak shares many basic principles of logic programming, such as term unification and sequential execution of queries, a basic interpreter can be obtained by rewriting the derivation rules of the operational semantics as clauses of a logic program. To this end, consider the following example encoding of the reactive agent program from above:

```
adjacent(a,b).
adjacent(b,c).
at(agent,a).
at(gold,a).
at(depot,c).
```

```
do(pick(gold)) :- at(agent,X),
                  retract(at(gold,X)),
                  assert(carries(agent,gold)).
do(move(X,Y)) :- at(agent,X),
                  retract(at(agent,X)),
                  assert(at(agent,Y)).
do(drop(gold)) :- at(agent,X),
                  retract(carries(agent,gold)).

procedure(+(at(gold,X)),P) :-
   at(agent,X), at(depot,Y),
   P = [do(pick(gold)), !(at(agent,Y)), do(drop(gold))].

procedure(+(!(at(agent,X))),P) :- at(agent,X), P=[].

procedure(+(!(at(agent,X))),P) :-
   at(agent,Y), not X=Y, adjacent(Y,Z), not at(obstacle,Z),
   P = [do(move(Y,Z)), !(at(agent,X))].
```

The beliefs are directly specified as facts, which are dynamically changed when an action is executed that affects some of the beliefs. In accordance with Definition 6.2.2, the specification of a procedure consists of a triggering event, conditions on the beliefs, and a sequence of actions or goals.

Based on an AgentSpeak program, which contains an initial belief base along with a specification of the actions and procedure definitions, a logic program that acts as an interpreter can be obtained by a direct translation of the derivation rules from above; see Figure 6.3. The

```
desire(E,I) :- procedure(E,P), intention([P|I]).

intention([[do(A)|P]|I]) :- do(A), intention([P|I]).

intention([[!(F)|P]|I]) :- desire(+(!(F)),[P|I]).

intention([[?(F)|P]|I]) :- F, intention([P|I]).

intention([[?(F)|P]|I]) :- desire(+(?(F)),[P|I]).

intention([[]|I]) :- intention(I).

intention([]).
```

FIGURE 6.3: A generic AgentSpeak interpreter.

only restriction in comparison to the general operational semantics is that desires cannot be dropped and that desires and intentions are selected in the order in which they arise. This does not allow us to delay the adoption of a desire in case a relevant but no currently applicable procedure exists. The interpreter does not include a definition for sensing, which requires the addition of an interface to the execution body of the agent with its sensors.

Consider, as an example, the query `desire(+(at(gold,a)),[])`. It admits a successful derivation, in which the intention of the agent evolves as follows:

```
[[do(pick(gold)), !(at(agent,c)), do(drop(gold))]]
[[!(at(agent,c)), do(drop(gold))]]
[[do(move(a,b)), !(at(agent,c))], [do(drop(gold))]]
[[!(at(agent,c))], [do(drop(gold))]]
[[do(move(b,c)), !(at(agent,c))], [], [do(drop(gold))]]
[[!(at(agent,c))], [], [do(drop(gold))]]
[[], [], [], [do(drop(gold))]]
[[], [], [do(drop(gold))]]
[[], [do(drop(gold))]]
[[do(drop(gold))]]
[[]]
[]
```

6.4   SPARK

The language and system SPARK (an acronym for the Stanford Research Institute Procedural Agent Realization Kit) has been developed for large-scale, practical applications of reactive agent programs. It, too, builds on the PRS-architecture and the underlying BDI-model for behavior-driven agents. SPARK, in comparison with AgentSpeak, provides more expressive means for encoding and controlling agents in rich and dynamic domains.

6.4.1   Syntax

The world model of a SPARK agent is based on a domain-specific set of belief atoms but may contain negative beliefs, too. A *belief literal* is a belief atom or its negation. At any state, a belief base consists of a finite set of variable-free belief literals. A negated literal is true only if it occurs explicitly in the belief base, which is in contrast to AgentSpeak, where negation-as-failure is applied to evaluate negated conditions against a belief base of atoms. A *belief formula* is built from the belief atoms and the standard logical connectives.

A domain signature in SPARK includes a set of action symbols to form *actions*. These represent all primitive actions that an agent can directly perform in its environment, but also

Task	Meaning
noop	do nothing
fail	fail
conclude φ	add the fact to the beliefs
retract φ	delete the fact from the beliefs
do a	perform the action
achieve φ	attempt to achieve φ
seq(τ_1, τ_2)	execute τ_1 then τ_2
if$(\varphi, \tau_1, \tau_2)$	if φ is true, execute τ_1, else τ_2
try(τ, τ_1, τ_2)	if τ succeeds, execute τ_1, else τ_2
wait(φ, τ)	wait until φ is true, then execute τ
while$(\varphi, \tau_1, \tau_2)$	repeat τ_1 until φ has no solution, then execute τ_2

FIGURE 6.4: The basic and compound elements of SPARK task descriptions. The expressions φ and a denote an arbitrary belief literal and action, respectively. The sub-tasks τ, τ_1, τ_2 are recursively defined using all available programming constructs.

abstract names for more complex behaviors. The desire to perform an action a, be it primitive or not, is always expressed in the belief base using the special belief atom $Desire(a)$. Further pre-defined belief expressions are $Success(a)$ and $Fail(a)$, and similarly $Desire(\varphi)$, $Success(\varphi)$, and $Fail(\varphi)$ for belief literals φ. *Triggers* for procedures in SPARK are of the form $Do(a)$, where a is a non-primitive action; $Achieve(\varphi)$, where φ is a belief literal whose achievement is desired; or $+\varphi$, indicating that the agent just came to believe literal φ.

SPARK supports a rich programming language for the specification of complex behaviors, so-called *task descriptions*; see Figure 6.4. A *procedure* in SPARK is then defined as an expression

$$e : \phi \leftarrow \tau$$

where e is a trigger, ϕ (the applicability condition) a belief formula, and τ a task description.

As a small example, consider these two procedures for a software agent that filters incoming messages:

$$Do(ForwardMessage(x)) : \neg IsSpam(x) \leftarrow \textbf{try } (\textbf{achieve } ClassifyMessage(x, y),$$
$$\textbf{do } AddToFolder(x, y),$$
$$\textbf{do } Forward(x)$$
$$) \tag{6.3}$$

$$Do(ForwardMessage(x)) : IsSpam(x) \leftarrow \textbf{do } Delete(x)$$

Put in words, if the non-primitive action *ForwardMessage*(x) is desired and x is not believed to be spam, then the agent tries to classify the message to be about some topic y. If this succeeds, the message is added to a folder named y, otherwise it is forwarded to the user. If message x is believed to be spam, it is simply deleted.

6.4.2 Operational Semantics

The operational semantics for SPARK can be given by a set of derivation rules in a similar fashion as in the case of AgentSpeak. The main difference between the two languages lies in the expressive programming language for procedures. Their semantics is defined in SPARK using the concept of a finite state machine.

State Transitions

The operational meaning of a complex task definition is described by a state transition system. Each such automaton has an initial state, denoted by s_0, and at least one, possibly both, of two distinct terminal states: s^+, the *success state*, and s^-, the *failure state*.

A single state transition is labeled with conditions for its applicability along with its effects on the knowledge base. A condition is a sequence $c_1, \ldots, c_n$ of expressions $c_i = \varphi$ or $c_i = \overline{\varphi}$, where φ requires this belief literal to be entailed by the belief base while $\overline{\varphi}$ requires that this is not the case. Thus there is a distinction between the absence of a belief, say, $\overline{Obstacle(B)}$, and the presence of the explicitly negated belief $\neg Obstacle(B)$. The effect of a state transition is a sequence $e_1, \ldots, e_m$ of expressions $e_i = \varphi$ or $e_i = \overline{\varphi}$ where φ can be any belief atom.

A state transition with conditions $\vec{c}$ and effects $\vec{e}$ is denoted by $s \xrightarrow{\vec{c} \,|\, \vec{e}} s'$. Empty conditions and effects are simply omitted. The construction of a finite state machine $\mathcal{M}(\tau)$ for an arbitrary, complex task description τ is defined recursively through the construction of a state machine for each task expression. For the basic tasks, the construction is as follows:

$$\mathcal{M}(\textbf{noop}) \stackrel{\text{def}}{=} \{s_0 \longrightarrow s^+\}$$

$$\mathcal{M}(\textbf{fail}) \stackrel{\text{def}}{=} \{s_0 \longrightarrow s^-\}$$

$$\mathcal{M}(\textbf{conclude } \varphi) \stackrel{\text{def}}{=} \{s_0 \xrightarrow{|\varphi} s^+\}$$

$$\mathcal{M}(\textbf{retract } \varphi) \stackrel{\text{def}}{=} \{s_0 \xrightarrow{|\overline{\varphi}} s^+\}$$

$$\mathcal{M}(\textbf{do } a) \stackrel{\text{def}}{=} \{s_0 \xrightarrow{|Desire(a)} s, s \xrightarrow{Success(a)|} s^+, s \xrightarrow{Fail(a)|} s^-\}$$

$$\mathcal{M}(\textbf{achieve } \varphi) \stackrel{\text{def}}{=} \{s_0 \xrightarrow{\varphi|} s^+, s_0 \xrightarrow{\overline{\varphi}|Desire(\varphi)} s, s \xrightarrow{Success(\varphi)|} s^+, s \xrightarrow{Fail(\varphi)|} s^-\}$$

According to this definition, the task to do an action is executed by adding the corresponding desire. It succeeds if at some point *Success*(a) occurs in the belief base and fails if eventually *Fail*(a) is believed. The task to achieve something succeeds immediately if the property happens to be true, otherwise a corresponding desire is added and its success or failure determines the final state transition.

The finite state machine for a compound task expression is obtained by recursively combining the individual state machines for each component, thereby applying appropriate substitutions of states and, in some cases, adding conditions to some of the state transitions. The latter is denoted by $c \Rightarrow \mathcal{M}(\tau)$, meaning that condition s is added to the sequence of conditions for all transitions in $\mathcal{M}(\tau)$ leading away from s_0.

$$\mathcal{M}(\mathbf{seq}(\tau_1, \tau_2)) \stackrel{\text{def}}{=} \mathcal{M}(\tau_1)\{s^+/s\} \cup \mathcal{M}(\tau_2)\{s_0/s\}$$
$$\mathcal{M}(\mathbf{if}(\varphi, \tau_1, \tau_2)) \stackrel{\text{def}}{=} \varphi \Rightarrow \mathcal{M}(\tau_1) \cup \overline{\varphi} \Rightarrow \mathcal{M}(\tau_2)$$
$$\mathcal{M}(\mathbf{try}(\tau, \tau_1, \tau_2)) \stackrel{\text{def}}{=} \mathcal{M}(\tau)\{s^+/s_1, s^-/s_2\} \cup \mathcal{M}(\tau_1)\{s_0/s_1\} \cup \mathcal{M}(\tau_2)\{s_0/s_2\}$$
$$\mathcal{M}(\mathbf{wait}(\varphi, \tau)) \stackrel{\text{def}}{=} \{s_0 {\longrightarrow} s\} \cup \varphi \Rightarrow \mathcal{M}(\tau)\{s_0/s\}$$
$$\mathcal{M}(\mathbf{while}(\varphi, \tau_1, \tau_2)) \stackrel{\text{def}}{=} \{s_0 {\longrightarrow} s\} \cup \varphi \Rightarrow \mathcal{M}(\tau_1)\{s_0/s, s^+/s\} \cup \overline{\varphi} \Rightarrow \mathcal{M}(\tau_2)\{s_0/s\}$$

Put in words, the sequential execution of two tasks corresponds to the sequential combination of the individual state machines in such a way that the second task can only be started if the first one ends in the success state, and then the success of the second one determines whether the entire sequence is successful. A conditional is modeled by adding the corresponding condition to the tasks representing the two cases. For the statement $\mathbf{try}(\tau, \tau_1, \tau_2)$, the initial state of τ_1 is identified with the success state of τ while the initial state of τ_2 is identified with the failure state of τ. Waiting for a condition before executing a task is modeled by a state transition to an internal state after which the task execution is possible only if the condition holds. Finally, the loop requires an initial state transition to an internal state s to which the execution of the repeated task returns until the condition is no longer satisfied, and then the exiting task is executed from s. For illustration, Figure 6.5 depicts the two finite state machines that constitute the operational semantics of the bodies of the two example procedures in (6.3).

Derivation Rules

With the help of the finite state machines as models for complex task descriptions, the operational semantics of SPARK can be given by a set of derivation rules which are similar to those in AgentSpeak. At any time during the execution of a SPARK program, the agent is characterized by a 4-tuple $(\mathcal{B}, \mathcal{D}, \mathcal{I}, \sigma)$ again, where $\sigma \in \{\mathtt{Sense}, \mathtt{Select}, \mathtt{Act}\}$ denotes the current stage in the sense-select-act cycle as before. In SPARK, the belief base $\mathcal{B}$ is a set of belief literals; the desires $\mathcal{D}$ are of the form $Desire(a)$, $Desire(\varphi)$, or $+\varphi$; and each intention in $\mathcal{I}$ is a tuple $\langle d, S, s \rangle$ consisting of a desire d, an instance S of a finite state machine, and the current local state s in S. The initial configuration is $(\mathcal{B}, \{\}, \{\}, \mathtt{Sense})$ with $\mathcal{B}$ being an arbitrary initial set of belief literals.

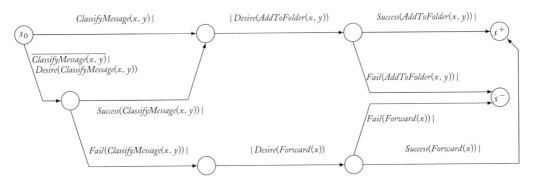

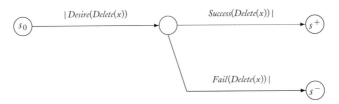

FIGURE 6.5: Finite state machines for two example procedures.

As in AgentSpeak, there are three selection functions:

- $\mathcal{S}_D$ selects an element from the current desires;
- $\mathcal{S}_P$ selects an applicable procedure for a desire;
- $\mathcal{S}_I$ selects an element $\langle d, S, s_1 \rangle \in \mathcal{I}$ with an allowed state transition $s_1 \xrightarrow{\vec{c} \,|\, \vec{e}} s_2$.

The only derivation rule for the stage in which the agent senses is,

$$\frac{(\mathcal{B}, \mathcal{D}, \mathcal{I}, \mathsf{Sense})}{(\mathcal{B}', \mathcal{D}', \mathcal{I}, \mathsf{Select})}$$

where $\mathcal{B}'$ is obtained by updating $\mathcal{B}$ according to the sensing result and where $\mathcal{D}$ is updated to $\mathcal{D}'$ by adding all desires thus sensed.

For the selection of a desire, three cases are distinguished. If there are no desires, the agent proceeds by selecting an intention. If the selected desire is of the form $Desire(x)$, then an applicable procedure is selected and added to the intentions if one exists, otherwise a "failure machine" is added to the intentions. If the selected desire is of the form $+\varphi$, then *all* applicable procedures are added to the intentions.

1. If $\mathcal{D} = \{\}$, then

$$\frac{(\mathcal{B}, \{\}, \mathcal{I}, \texttt{Select})}{(\mathcal{B}, \{\}, \mathcal{I}, \texttt{Act})}.$$

2. Suppose $\mathcal{S}_D(\mathcal{D}) = \textit{Desire}(x)$, where x is action a or belief literal φ.

    (a) If there is a procedure for $\textit{Do}(a)$ or $\textit{Achieve}(\varphi)$, respectively, that is applicable w.r.t. $\mathcal{B}$, then

$$\frac{(\mathcal{B}, \mathcal{D}, \mathcal{I}, \texttt{Select})}{(\mathcal{B}, \mathcal{D} \setminus \{\textit{Desire}(x)\}, \mathcal{I} \cup \{\langle \textit{Desire}(x), S, s_0 \rangle\}, \texttt{Act})}$$

    where S is the state machine for the task of the selected procedure instance $\mathcal{S}_P(\textit{Desire}(x))$.

    (b) If there are no such procedures that are applicable w.r.t. $\mathcal{B}$, then

$$\frac{(\mathcal{B}, \mathcal{D}, \mathcal{I}, \texttt{Select})}{(\mathcal{B}, \mathcal{D} \setminus \{\textit{Desire}(x)\}, \mathcal{I} \cup \{\langle \textit{Desire}(x), S, s_0 \rangle\}, \texttt{Act})}$$

    where $S = \{s_0 \overset{|\textit{Fail}(x)}{\longrightarrow} s^+\}$.

3. Suppose $\mathcal{S}_D(\mathcal{D}) = +\varphi$.

    (a) If there is a procedure for $+\varphi$ that is applicable w.r.t. $\mathcal{B}$, then

$$\frac{(\mathcal{B}, \mathcal{D}, \mathcal{I}, \texttt{Select})}{(\mathcal{B}, \mathcal{D} \setminus \{+\varphi\}, \mathcal{I} \cup \mathcal{I}', \texttt{Act})}$$

    where $\mathcal{I}'$ is the set of triples $\langle +\varphi, S, s_0 \rangle$ for all procedure instances applicable to $+\varphi$ w.r.t. $\mathcal{B}$.

    (b) If there are no procedures for $+\varphi$ that are applicable w.r.t. $\mathcal{B}$, then

$$\frac{(\mathcal{B}, \mathcal{D}, \mathcal{I}, \texttt{Select})}{(\mathcal{B}, \mathcal{D} \setminus \{+\varphi\}, \mathcal{I}, \texttt{Select})}.$$

For the last step in the sense-select-act cycle, if there are no intentions at all or no applicable state transition in any of the current intentions, then nothing changes and the agent simply starts the cycle again. Otherwise the next step in the selected behavior is executed. In case this state transition happens to end in one of the two possible terminal states, the intention is removed and, if it was for some $\textit{Desire}(x)$, the outcome (success or failure) is added to the belief base.

1. If $\mathcal{I} = \{\}$, then

$$\frac{(\mathcal{B}, \mathcal{D}, \{\}, \texttt{Act})}{(\mathcal{B}, \mathcal{D}, \{\}, \texttt{Sense})}.$$

2. Let $\mathcal{S}_I(\mathcal{I})$ be $\langle d, S, s_1 \rangle$ along with the transition step $s_1 \xrightarrow{\vec{c}\,|\,\vec{e}} s_2$.

(a) If $s_2 \notin \{s^+, s^-\}$, then

$$\frac{(\mathcal{B}, \mathcal{D}, \mathcal{I}, \mathsf{Act})}{(\mathcal{B}', \mathcal{D}', \mathcal{I} \setminus \{\langle d, S, s_1 \rangle\} \cup \{\langle d, S, s_2 \rangle\}, \mathsf{Sense})};$$

(b) If $d = +\varphi$ and $s_2 \in \{s^+, s^-\}$, then

$$\frac{(\mathcal{B}, \mathcal{D}, \mathcal{I}, \mathsf{Act})}{(\mathcal{B}', \mathcal{D}', \mathcal{I} \setminus \{\langle d, S, s_1 \rangle\}, \mathsf{Sense})};$$

(c) If $d = Desire(x)$ and $s_2 = s^+$, then

$$\frac{(\mathcal{B}, \mathcal{D}, \mathcal{I}, \mathsf{Act})}{(\mathcal{B}' \cup \{Success(x)\}, \mathcal{D}', \mathcal{I} \setminus \{\langle d, S, s_1 \rangle\}, \mathsf{Sense})};$$

(d) If $d = Desire(x)$ and $s_2 = s^-$, then

$$\frac{(\mathcal{B}, \mathcal{D}, \mathcal{I}, \mathsf{Act})}{(\mathcal{B}' \cup \{Fail(x)\}, \mathcal{D}', \mathcal{I} \setminus \{\langle d, S, s_1 \rangle\}, \mathsf{Sense})}.$$

In each case, $\mathcal{B}'$ and $\mathcal{D}'$ are the results of updating $\mathcal{B}$ and $\mathcal{D}$, respectively, by the effects $\vec{e}$ of the state transition.

This completes the definition of the operational semantics for SPARK programs. Available implementations of this language include several features that have not been treated in this introduction, such as the parallel execution of sub-tasks or the use of meta-beliefs and meta-procedures to modify the general behavioral stance of an agent.

6.5 EXERCISES

6.1. Extend the example AgentSpeak procedures from Section 6.2 to a complete program for a gold mining agent! Extend the generic interpreter by an interface to some agent executor and write a simulator to test the program with different scenarios similar to the one shown in Figure 5.5!

6.2. Specify the three AgentSpeak procedures from Section 6.2 in SPARK! Construct the corresponding finite state machines and find a complete derivation starting with the same initial beliefs as in Figure 6.2!

6.3. Extend the domain from Exercise 6.1 to allow for both dynamic as well as static obstacles, whose locations the agent should memorize! Assume a multiagent setting, where four agents collaborate in a team and write four different SPARK programs for these agents such that each team member takes a different role, for example explorer or collector!

Suggested Further Reading

PROCEDURAL ACTION PROGRAMS

1. Michael Genesereth, Nathaniel Love, and Barney Pell. General game playing. *AI Magazine* 26(2):73–84, 2006.
2. Giuseppe De Giacomo, Yves Lespérance, and Hector Levesque. ConGolog, a concurrent programming language based on the situation calculus. *Artificial Intelligence* 121(1–2):109–169, 2000.
3. Giuseppe De Giacomo, Yves Lespérance, Hector Levesque, and Sebastian Sardiña. On the semantics of deliberation in IndiGolog—from theory to practice. In *Proceedings of the International Conference on Principles of Knowledge Representation and Reasoning*, pages 603–614, Toulouse, France, 2002.
4. Erik Mueller. *Commonsense Reasoning*. Morgan Kaufmann 2006.
5. Raymond Reiter. *Knowledge in Action*. MIT Press 2001.
6. Murray Shanahan. *Solving the Frame Problem: A Mathematical Investigation of the Common Sense Law of Inertia*. MIT Press 1997.

PLANNING

1. Fahiem Bacchus and Froduald Kabanza. Using temporal logic to express search control knowledge for planning. *Artificial Intelligence* 116(1–2):123–191, 2000.
2. Meghyn Bienvenu and Sheila McIlraith. Planning with qualitative temporal preferences. In *Proceedings of the International Conference on Principles of Knowledge Representation and Reasoning*, pages 134–144, Lake District, UK, 2006.
3. Jonas Kvarnström and Patrick Doherty. TALplanner: a temporal logic based forward chaining planner. *Annals of Mathematics and Artificial Intelligence* 30(1–4):119–169, 2000.

DECLARATIVE ACTION PROGRAMS

1. Thom Frühwirth. Theory and practice of constraint handling rules. *Journal of Logic Programming* 37(1–3):95–138, 1998.
2. Michael Thielscher. *Reasoning Robots: The Art and Science of Programming Robotic Agents*. Kluwer 2005.

REACTIVE ACTION PROGRAMS

1. Rafael Bordini, Jomi Hübner, and Michael Wooldridge. *Programming Multi-Agent Systems in AgentSpeak using Jason*. Wiley 2007.

2. Viviana Mascardi, Daniela Demergasso, and Davide Ancona. Languages for programming BDI-style agents: an overview. In *Proceedings of the Workshop From Objects to Agents*, pages 9–15, Camerino, Italy, 2005.

3. SPARK Reference Manual. `www.ai.sri.com/~spark`.

References

Bratko, I., *Prolog Programming for Artificial Intelligence.* 3rd edn., Reading, MA: Addison-Wesley, 2000.

Clocksin, W. F. and Mellish, C. S., *Programming in PROLOG: Using the ISO Standard.* Berlin: Springer, 2003.

Genesereth, M., Love, N. and Pell, B., "General game playing," *AI Magazine*, 26(2):73–84, 2006.

Georgeff, M. P. and Lansky, A. L., "Reactive reasoning and planning," in *Proceedings of the AAAI National Conference on Artificial Intelligence*, Seattle, pp. 677–682, 1987.

Hindriks, K., De Boer, F., Van der Hoek, W. and Meyer, J.-J., "Agent programming in 3APL," *Autonomous Agents and Multi-Agent Systems*, 2(4):357–401, 1999. doi:10.1023/A:1010084620690

Kowalski, R. and Sergot, M., "A logic based calculus of events," *New Generation Computing*, 4:67–95, 1986.

Levesque, H., Reiter, R., Lespérance, Y., Lin, F. and Scherl, R., "GOLOG: A logic programming language for dynamic domains," *Journal of Logic Programming*, 31(1–3):59–83, 1997. doi:10.1016/S0743-1066(96)00121-5

McCarthy, J., "Programs with Common Sense," in *Proceedings of the Symposium on the Mechanization of Thought Processes*, London, vol. 1, pp. 77–84, November 1958.

McCarthy, J., *Situations and Actions and Causal Laws.* Stanford Artificial Intelligence Project, Memo 2, Stanford University, CA, 1963.

McCarthy, J. and Hayes, P. J., "Some philosophical problems from the standpoint of artificial intelligence," *Machine Intelligence*, 4:463–502, 1969.

McDermott, D., "The 1998 AI planning systems competition," *AI Magazine*, 21(2):35–55, 2000.

Morley, D. and Meyers, K., "The SPARK agent framework," in *Proceedings of the International Conference on Autonomous Agents and Multiagent Systems*, pp. 714–721, 2004.

Nilsson, N. J., *Artificial Intelligence: A New Synthesis.* San Mateo, CA: Morgan Kaufmann, 1998.

Rao, A., "AgentSpeak(L): BDI agents speak out in a logical language," in W. Van de Velde and J. Perram, eds., *Agents Breaking Away, LNAI*, Berlin: Springer, vol. 1038, pp. 42–55, 1996. doi:10.1007/BFb0031845

Russell, S. and Norvig, P., *Artificial Intelligence: A Modern Approach*. Englewood Cliffs, NJ: Prentice-Hall, 2nd edn., 2003.

Shanahan, M., *Solving the Frame Problem: A Mathematical Investigation of the Common Sense Law of Inertia*. Cambridge, MA: MIT Press, 1997.

Thielscher, M., "From situation calculus to fluent calculus: State update axioms as a solution to the inferential frame problem," *Artificial Intelligence*, 111(1–2):277–299, 1999. doi:10.1016/S0004-3702(99)00033-8

Thielscher, M., "FLUX: A logic programming method for reasoning agents," *Theory and Practice of Logic Programming*, 5(4–5):533–565, 2005. doi:10.1017/S1471068405002358

Author Biography

Michael Thielscher is a Professor and head of the Computational Logic Group at Dresden University in Germany since 1997. He received his PhD in Computer Science from Darmstadt University of Technology, Germany. His research is mainly in Knowledge Representation, Cognitive Robotics, Commonsense Reasoning, Game Playing, and Constraint Logic Programming. He has developed the action programming language and system FLUX and has published numerous papers and two books on knowledge representation for actions, on comparisons of different action languages, and on implementations of action programming systems. In 1998, his Habilitation thesis was honored with the award for research excellence by the alumni of Darmstadt University of Technology. He co-authored the program FLUXPLAYER, which in 2006 was crowned the world champion at the Second General Game Playing Competition in Boston.

Printed in the United States
by Baker & Taylor Publisher Services